QUINOA

(*Chenopodium quinoa Willd.*)

Other Related Books

Title	Author	ISBN
• Applied Microbiology (Agriculture, Environmental, Food and Industrial Microbiology)	*S.M. Reddy*	9789386237903
• Arid Fruit Research	*A. Chandra*	9788172330262
• Beverages: Processing and Technology	*Deepak Mudgil*	9789387869363
• Biopesticides in Environment and Food Security: Issues and Strategies	*Opender Koul*	9788172337971
• Functional Foods: Sources and Health Benefits	*Deepak Mudgil*	9789386237002
• Herbaceous Plants as Natural Protective Food	*M.K. Rana*	9788172339005
• Horticulture and Livelihood Security	*Prem Nath*	9788172336745
• Objective Food Science & Technology 3rd Ed (Forthcoming)	*Deepak Mudgil*	9789388172868
• Phenolic Antioxidants and Health Benefits	*U.D. Chavan*	9789386652133
• Sustainable Horticulture Development and Nutrition Security	*Prem Nath*	9789386652386
• Tamarind Science and Technology	*A. Balasubramanian*	9789386652256
• The Basics of Human Civilization Food, Agricutlure and Humanity (Vol. 3): Agricutlure	*Prem Nath*	9788172338909
• Vegetables and Their Allied as Protective Food	*M.K. Rana*	9788172338992

QUINOA

(*Chenopodium quinoa Willd.*)

A potential crop for future food, health security, livelihood generation and poverty eradication

Dr. Dharm Singh

M.Sc (Ag)., M.H.Sc. (London), Ph.D., D.Sc., FASA (USA), MISHS (Netherland)

Director General

Indian Medicinal Plants Marketing Federation

(Federal Union for Research & Development of Medicinal Plants in India)

Member

International Bee Research Association (UK), International Society for Horticultural Sciences, Netherland & Fellow of Royal Horticulture Society, London (UK)

SCIENTIFIC PUBLISHERS

Published by
SCIENTIFIC PUBLISHERS (INDIA)
Jodhpur –
5 A, New Pali Road
P.O. Box 91
Jodhpur - 342 001 INDIA

ISBN: 978-93-87893-98-6
E-ISBN: 978-93-87991-67-5

Visit the Scientific Publishers (India) website at
http://www.scientificpub.com
Printed in India

(*Late Smt. Sarvati Devi*)

DEDICATION

My informal Mother-teacher "Late Smt. Sarvati Devi", a noted social worker of Agra district, Uttar Pradesh, insisted me to work on quinoa, a nutrient-rich poor men's food, to ensure its supply and availability to resource poor people of remote areas. She dedicated 42 years of her life in socio-economic upliftment and health improvement of mal-nourished mothers, pregnant women, children and poor people. She arranged seeds of quinoa in 1975 from South America and sown in two different villages of Bichpuri Block of Agra i.e., Chhontana and Nagla Patiram during Rabi season of 1975–76. This was the first attempt of quinoa introduction in India, but, un-luckily, she not get complete success due to lack of desired technical knowhow and suitable growing protocol. After this, nobody has taken further initiation for quinoa development in India. Thus, inspite of nutrient rich super food and wide adoptability, quinoa was remain un-explored till 2003.

After 30 years of first attempt, I started a pilot testing and extension of wonderful quinoa during 2004–05 in different states of India. It was the first time in India when quinoa was successfully grown & well-established in different states. The full credit of this second successful introduction in real sense gone to my divine Mother-teacher Late Smt. Sarvati Devi, who made first attempt for quinoa introduction during 1975-76. Therefore, it is my moral duty to dedicate this valuable manuscript to beloved mother of "Rural Poor People" who devoted her entire life in the service of downtrodden, neglected and unserviced communities of the rural nation. I feel proud to inscribe this text to Late Smt Sarvati Devi to honour for her momentous, unvalued and unforgettable long sacrifice for the people of India.

Quinoa (*Chenopodium quinoa Willd*) - A wonder on the earth

राधा मोहन सिंह
RADHA MOHAN SINGH

सत्यमेव जयते

कृषि एवं किसान कल्याण मंत्री
भारत सरकार
MINISTER OF AGRICULTURE
& FARMERS WELFARE
GOVERNMENT OF INDIA

16, April, 2018

D.O. No. 942 /AM

MESSAGE

Agriculture is the mainstay of Indian economy, as it constitutes the backbone of the rural livelihood security system. It is the core of planned economic development in India as the trickle-down effect of agriculture is significant in reducing poverty and regional inequality in the country. Traditional crops like wheat, rice and other cereals are the mostly used staple foods by the majority of the population, the availability and supply of which gets affected due to biotic and abiotic factors. This situation needs an alternative food crops having full nutrition and strong enough to withstand to abiotic and biotic stresses.

Quinoa could be an important nutritive food of the future and meet the challenges of hunger led malnutrition as it has better adaptability to salinity, drought and climate change. Narratively, it can be termed as super food on the platter highly rich in digestive protein, essential fatty acids, minerals, vitamins and dietary fiber.

Information on quinoa and its production technology could be of immense use to many of the interested individual and organizations for its popularization, commercial production and processing. I am confident that his book would be very useful to food technologists, industrialists, researchers and policy makers of varied fields especially agriculture, health and nutrition.

I congratulate Indian Medicinal Plants Marketing Federation, for his commendable work towards compilation and publication of this valuable reference book.

Radha Mohan Singh

(Radha Mohan Singh)

Dr. S.K.PATTANAYAK
SECRETARY

सत्यमेव जयते

भारत सरकार
कृषि एवं किसान कल्याण मंत्रालय
कृषि, सहकारिता एवं किसान कल्याण विभाग
Government of India
Ministry of Agriculture & Farmers Welfare
Department of Agriculture, Cooperation & Farmers Welfare

FOREWORD

Agriculture is not only a food producing enterprise but a key driver of development process forming the backbone of livelihood system in India. Agriculture production will continue to be a high priority objective for food security, livelihood security, natural resource security, industrial development and poverty alleviation. Impact of climate change and resource degradation has recently affected this sector which is likely to impact production of traditional food crops. Climate change may erode the productivity of the country by 4 to 6% of GDP. Traditional crops like wheat, rice and maize are likely to be adversely affected due to a number of biotic and abiotic factors, water scarcity, soil salinity, water logging, floods and poor soil nutrition.

In this perspective, quinoa may be one of the alternate crop to support national food security and silver bullet to address the major issues of health security and livelihood generation. Quinoa, a plant of the Amaranthaceae family, also known as goose foot is largely found in Andes. Quinoa is not a cereal grain crop technically but instead a pseudo-cereal as it is used similar to cereals. It is highly resistant and adaptable crop to wide range of climatic conditions, salinity and drought and most resilient to climate change induced weather aberration, resulting in no risk of crop failure under extreme agro climatic conditions. Besides, quinoa has the ability to uplift small and marginal farmers from poverty and can provide economic growth to them. In spite of vast potential in providing food security, health security, livelihood security and better opportunities to the development of food Industries, quinoa remains un-familiar to majority of the people and its commercial potential and benefits are still untapped in India.

I am happy to observe that the efforts made by Dr. Dharm Singh, Director General, Indian Medicinal Plants Marketing Federation in compilation, publication and popularization of quinoa as a wonderfull super food crop deserves praise.

I congratulate the author for his devotion and sustained efforts in writing this valuable text which may be useful for food technologists, industrialists, line-departments, research organizations, scientists, dieticians, consumers and farmers of this country.

(S.K.Pattanayak)

Date: 6th April, 2018

प्रो. आशुतोष शर्मा
Prof. Ashutosh Sharma

सत्यमेव जयते

सचिव
भारत सरकार
विज्ञान और प्रौद्योगिकी मंत्रालय
विज्ञान और प्रौद्योगिकी विभाग
Secretary
Government of India
Ministry of Science and Technology
Department of Science and Technology

FOREWORD

Agriculture is not only the food producing machine but key driver of development process and backbone of livelihood system in India. Agriculture production will continue to be a high priority objective for Food security, livelihood security, natural resource security, industrial development and poverty alleviation. Impact of climate change and resource degradation has recently hit this sector most violently resulting stagnation / downfall in production of traditional food crops leading to fear of high food in-security and malnutrition. India is home to the largest malnourished population in the world where 194.6 million people go hungry every day, 30.7% of children are under weight, 38.7% are stunted and 3000 children die every day from poor diet related illness. These problems eroding the productivity of the country by 4 to 6% of GDP. Traditional crops like wheat, rice and maize are at stake due to a number of biotic and a biotic factors, water scarcity, soil salinity, water logging, floods and poor soil nutrition.

Quinoa *(Chenopodium quinoa Willd)* is acknowledged as one of the most important "Functional foods" for long and healthy human life. It is highly nutrient rich, well-recognized and established food for vegetarians which could address both food and health security problems in India. Also, it can alleviate malnutrition, promote food industries, generate employment, uplift Farm-output and meet the nutritional requirement of poor people, diabetics, cardiac patients and mal-nourished children. Quinoa grains, containing dietary fiber, reduces the risk of high blood pressure and heart attack. Due to its climate hardy nature and very low water requirement, quinoa has the highest potential to produce food under climate change & harsh environments, drought, salinity and water scarcity conditions. In the present trend of healthy foods, quinoa attracted people world over and attained top position among nutrient-rich foods. Thus, it can be a "Silver bullet" for economic upliftment of rural people and major contributor to national agendas for future food and nutritional security, if quinoa is ommercialized and its wonderful benefits are tapped.

Despite of high adoptability under marginal environment, nutrient rich and medicinal values, quinoa remains unfamiliar and its commercial potential is still untapped. In this context, Dr. Dharm Singh, Director General, Indian Medicinal Plants Marketing Federation has done an appreciable, commendable & valuable work and compiled all relevant information on quinoa in the form of a reference book entitled "Quinoa-A potential crop for future food, health security, livelihood generation and poverty eradication". His efforts in this form will certainly help in commercialization of this un-explored wonder crop in India and help in creating the interest in quinoa production, processing, value-addition and development of food-based industries. This book has really a great importance as it contains recent literature and provides a valuable platform on complete knowledge on quinoa, a wonder super food crop on the earth.

I congratulate the learned author for his devotion, sustained efforts and valuable contribution.

(Ashutosh Sharma)

J. P. MEENA, IAS

सचिव
भारत सरकार
खाद्य प्रसंस्करण उद्योग मंत्रालय
Secretary
Government Of India
Ministry of Food Processing Industries

FOREWORD

Quinoa has gained recognition world over as one of the most important super foods which provides all the essential life sustaining nutrients. It is a grain crop grown primarily for its edible seeds to support nutritional requirements. Quinoa has ability to uplift economic conditions of small and marginal farmers as it is highly resilient to agro-ecological extremes (soil, temperature, rainfall & altitude) with high tolerance to salinity and drought and does not demand high soil fertility, nutrition, water and labor.

2. Quinoa is being seen as a wonder crop having potential to provide food security, health security, livelihood security and better opportunities for job creation in rural areas particularly for poor and economically weaker population across the globe. In spite of high nutrient status, high therapeutic value and wide adaptability under varied climatic situations, Quinoa remains un-familiar to majority of the people in India and its commercial potential and benefits are still untapped. Processing and value addition of Quinoa grains offer huge potential of creating large scale economic activities in the country.

3 In this context, the land mark work done by Dr. Dharm Singh, Director General, Indian Medicinal Plants Marketing Federation in compilation and publication of reference book on Quinoa will go a long way in awareness creation, popularization, conducting research and commercialization of this valuable unexplored crop in India.

4 I congratulate Dr. Dharm Singh, the learned author of this book, for his sustained efforts and useful contribution to popularizing Quinoa cultivation and its use for greater good of humanity.

(J.P. Meena)
03.04.2018

联合国
粮食及
农业组织

FOOD AND AGRICULTURE ORGANIZATION OF THE UNITED NATIONS

ORGANISATION DES NATIONS UNIES POUR L'ALIMENTATION ET L'AGRICULTURE

ORGANIZACION DE LAS NACIONES UNIDAS PARA LA AGRICULTURA Y LA ALIMENTACION

منظمة الأغذية والزراعة للأمم المتحدة

OFFICE OF THE REPRESENTATIVE IN INDIA

55, Lodi Estate New Delhi – 110 003

Tel: 24628877/24693060
Fax: 0091-11-24620115

Email: FAO-IN@fao.org

FOREWORD

While the share of agriculture in the Gross Domestic Product has decreased steadily in India over time, almost half of the population continues to depend upon agriculture for employment and income. Indian agriculture has grown significantly in last five decades and both crops and milk production show impressive achievements. Despite these achievements in agriculture rate of malnourishment continues to remain an area of concern. We can attribute this situation to a number of factors: Decreasing dietary diversity is one of them. In finding a solution to this problem, we need to be cognizant of the fact that a significant part of the Indian population is vegetarian by cultural tradition. This requires us to look for expanding the range of crops that are cultivated in India. A related concern in relation to the food and nutrition security is the impending climate change which will require us to look for crops that help us to respond to the dangers posed by threat multipliers such as drought. Quinoa is one such crop that helps us to address all these three concerns at the same time and is one of the best alternative to ensure sustained production, also in the context of increasing climate variability, and enhance food and nutrition security.

With high nutritive value and recognized as potential super food for present and future generations, Quinoa is a highly promising crop. It is un-demanding in agronomic terms, altitude-hardy, drought-resistant and highly adaptable to wide range of climate change. Salinity and drought affect this crop least and is highly tolerant to the climate change induced weather aberration. This helps in greatly reducing the risk of crop failure under water stress/ rain fed conditions. It can therefore be a cash crop for millions of farmers, providing raw material for food industries and nutritive food to people, especially those who are vegetarians. Quinoa can also reduce dependence on other staple like wheat and rice and help combatting hunger and conserving valuable natural resources like water. Quinoa can easily become a crop of choice in marginal areas of India where other major traditional food crops have so far have been dominant but are gradually failing due to climate change and degradation of soil and water resources. It has started attracting people world over and as a consequence there has been dramatic increase in demand in recent years. However, its nutritional and therapeutic benefits are still not fully realized in India owing to lack of authentic literature and general level of awareness.

In the above light, compilation of relevant literature on quinoa by Dr. Dharm Singh, Director-General, Indian Medicinal Plants Marketing Federation is really commendable, valuable and timely. I congratulate him for publication of this book, which may be a milestone for its readers and consumers in understanding, commercializing and utilizing this undervalued food crop.

Shyam Khadka
FAO Representative in India

27 April 2018

कृषि वैज्ञानिक चयन मंडल
(भारतीय कृषि अनुसंधान परिषद)
कृषि अनुसंधान भवन-I, पूसा, नई दिल्ली 110 012

AGRICULTURAL SCIENTISTS RECRUITMENT BOARD
(INDIAN COUNCIL OF AGRICULTURAL RESEARCH)
Krishi Anusandhan Bhavan I, Pusa, New Delhi 110 012

प्रो. (डॉ.) ए.के. श्रीवास्तव
अध्यक्ष

Prof. (Dr.) A.K. Srivastava
Chairman

सन्दर्भ सं/Ref. No. PPS/CHAIRMAN/2018
दिनांक/Dated 12.04.2018

FOREWORD

Quinoa is acknowledged as one of the most popular "Functional foods" for long and healthy human life. It is well-recognized and established as food for vegetarians, which provides most of the essential nutrients, gluten free protein, vitamins and minerals, and also with therapeutic attributes on skin and hair. Moreover, this food crop is extra-ordinarily adaptable to different agro-ecological zones, being grown in areas with relative humidity from 40% to 88%, from sea level to 2000 m above sea level and at temperatures from - 8^0 C to 38^0 C. It is a water efficient crop which is tolerant to the lack of soil moisture also. Thus, quinoa can be an excellent alternative food crop for India, in context of progressing climate change, harsh environment and limited availability of resources.

Despite of high nutrient and therapeutic values, and high adoptability under marginal environment, in India, quinoa remains un-familiar and its commercial potential is still untapped. In this endeavor, Dr. Dharm Singh, Director General, Indian Medicinal Plants Marketing Federation has done an appreciable job and compiled all the available useful information on quinoa in form a "Reference Book" entitled "Quinoa-A potential crop for future food, health security, livelihood generation and poverty eradication". This attempt will certainly help in popularization of this un-explored crop in India and help in creating the interest in quinoa production, processing and value-addition. This book has great importance as it reviews recent literatures and provides a valuable platform on complete knowledge on quinoa.

I congratulate the learned author for his valuable contribution.

Srivastava

(A.K. Srivastava)

FOREWORD

Agriculture is not only the food producing machine but key driver of development process and backbone of livelihood system in India. Agriculture production will continue to be a high priority objective for Food security, livelihood security, natural resource security, industrial development and poverty alleviation. Impact of climate change and resource degradation has recently hit this sector most violently resulting stagnation / downfall in production potential of traditional food crops leading to fear of high food in-security and malnutrition. India has globally highest hungry people that is 194.6 million, one-fourth of the world's population. These problems eroding the productivity of the country by 4 to 6% of GDP. Traditional crops like wheat, rice and maize are at stake due to a number of biotic and a biotic factors, water scarcity, soil salinity, water logging, floods and poor soil nutrition.

In this perspective, quinoa may be one of the alternate to support food security and could be the Silver bullet to address the major issues of health security and livelihood generation in India. It is highly resistant and adaptable crop to wide range of climate change, salinity and drought with least effect by other climate change induced weather aberration, resulting no risk of crop failure under water stress / rain fed conditions. Besides, quinoa has ability to uplift small & marginal farmers from poverty and can provide economic growth to them. In spite of vast potential in providing food security, health security, livelihood security and better opportunities to the development of food Industries, quinoa remains un-familiar to majority of the people and its commercial potential and benefits are still untapped in India.

In this exertion, efforts made by Dr. Dharm Singh, Director General, Indian Medicinal Plants Marketing Federation for publication of this knowledge resource book on Quinoa are excellent and forward looking in popularization of this valuable un-explored wonder super food crop.

I congratulate the author of this book for his devotion & sustained efforts for writing this valuable and timely text which may be useful for food technologists, industrialists, line-departments, research organizations, scientists, consumers etc.

Date: 21-03-2018

(Balraj Singh)
Vice Chancellor
Agriculture University, Jodhpur

PREFACE

India is a country of agrarian economy in which over 60 percent people presently depends on agriculture and allied activities. This sector is the principal source of livelihood for over 60 per cent of rural households. During 1965-66, wheat revolution or "Green revolution" made the people self-sufficient in food grains and India became the largest producer of wheat in the world. Due to emerging problems of global warming, impending climatic variability, drought and flood, water logging, soil salinity and nutrient deficiency in the soil often resulted crop failure, uncertainty of crop production and reduction in food grain output leading to its shortage and poor availability of food to the people. Under this situation, food security is a serious concern and major challenge to feed our fast growing population in India. Over 195 million people presently not getting adequate two square meals in a day.

On other hand, health sector remains un-prosperous in India resulting in a serious problem of malnutrition at house hold level due to lack of adequate food or inadequate nutritive food. India tops global charts in malnutrition and eroding the productivity of the country by 4 to 6 percent of GDP. Malnutrition is over 45 percent among children and 70 percent among pregnant women. This problem is becoming serious day by day and needs special attention to solve mal-nutrition and hunger-led diseases like diabetes & cardiac problems. In this context, we have reinvigorated efforts to search alternative crops rich in nutritional and biological values but strong enough to withstand to abiotic and biotic stresses.

Quinoa is identified as best alternative crop having better adaptability to drought and salinity with least effect by other climate change induced weather aberration. It is nutritionally, termed as "Super-food" by dietitians and nutritionists. Quinoa is also called "Golden grains", "Power food", "complete food" and "Mother of all grains". It is only the single food which can supply complete protein, all essential life sustaining nutrients and can reduce the risk of various diseases like blood cholesterol, blood pressure, diabetes, sexual weakness etc in very effective and preventive way. The main reasons of its selection as food for future are its wide adaptability to climate change and harsh environments, high nutritional value and its power as staple food of world's poorest people. Due to its major role in Food Security, Nutritional Security and Economic development, Quinoa is considered as Life, Hope, Past, Present and Future of the mankind on this planet. In 1996, quinoa was listed by FAO as one of humanity's most promising crop, not only for its beneficial properties and many uses, but also as an alternative to solve the serious problems of human nutrition.

Quinoa is highly resilient to agro-ecological extremes (soils, rainfall, temperature, altitude), tolerant to frost, drought and salinity and one of the most nutrient rich food in the world for human beings. National Aeronautics and Space Administration (NASA) included quinoa within the "Controlled Ecological Life

Support System" to equip its "Space craft" in long duration space travel as an ideal food-product for consumption during long-term space mission and whenever a crop would need to be grown in a space-ship. Quinoa has now been singled "high nutritive value food" on the earth with impressive biodiversity and an important role to play in the achievement of food security worldwide.

Quinoa is cholesterol and gluten free food having low and healthy fat as well as rich source of complete protein. It is also rich source of vitamin-B (niacin, riboflavin, thiamine etc.), Vitamin-E and dietary minerals (iron, magnesium, phosphorus, zinc). Lignin (Phyto-nutrient), a valuable nutrient to heart, is abundant in quinoa seeds. Quinoa grains, containing dietary fibre, reduces the risk of high blood pressure and heart attack. Its plants withstand temperatures from -8°C to 38°C and relative humidity from 40 percent to 88 percent. It is a highly water efficient plant, tolerant and resistant to lack of soil moisture and produces acceptable yields with rainfall of 100 to 200 mm. Quinoa, thus, could be an important food crop which can play a potential role in global food-security, eradication of malnutrition and poverty due to its higher nutritional qualities and agronomical versatility. It can be consumed like a staple food, replacing wheat or rice. Apart from its nutritive values, quinoa has the ability to uplift farmers from poverty. We need wide popularity of this potential crop and "Quinoa revolution" in India on top priority which could be the ''Silver bullet" to address major challenging issues of hunger, malnutrition, poverty and self-employment.

Until 2008, the production of quinoa was limited to Peru and Bolivia contributing 92 percent of total global quinoa production. Equator was another growing country with 4 percent contribution to global quinoa production. Remaining quinoa was produced by Chile and Northern Argentina. The first known experiment on quinoa conducted outside the Andean Region took place in 1935 in Kenya (obtained from Royal Botanical Garden) and secondly in England during 1970s. In India, it was first introduced in 1975-76 by Late Smt. Sarvati Devi, A great social worker of Agra (U.P.) and by author of this book during 2004-05. In other countries, quinoa was introduced to Denmark in 1984, Tibet in 1984, Netherland in 1986 and China in 1988. Currently, quinoa is in a process of expansion in non-domesticated countries and its cultivation is spreading rapidly throughout the world. After declaration of 2013 as "International Year of Quinoa" by UN, its demand and cultivation increased very fast. At present, it is grown in about 95 countries and total current production is 192.5 thousand metric tons (2014-15). In India, the area under quinoa cultivation during 2017-18 was 8630 hectares and production 20626 tons. Andhra Pradesh, Rajasthan, Uttar Pradesh, Maharashtra and Tamil Nadu are major quinoa growing states which occupies more than 55 percent of total area under quinoa cultivation.

Area of growing quinoa in Europe increased from 0 in 2008 to 5000 hectare in 2015, mainly in France, Spain and UK. It is largely cultivated in Bolivia, Peru and Equator in South America, United States of America, Australia, Japan, France, Italy and China. Quinoa recently attracted the people of developing Asian and European countries for using as food, feed, processing, value addition and industrial uses. India is in privileged position to promote quinoa production, quinoa-based food industries and its nutritional and cultural values with a view to eradicate hunger and malnutrition. In context to various health related problems and present trend of healthy foods, healthy environment and bio-based economy, quinoa attained top position to provide food security, health security and un-limited opportunities for the development of food and non-food industries.

For proper development of quinoa based food industries, conducting research projects and augment its production in India. Therefore, it is very important to provide scientific and authentic literature on various topics related to quinoa and create mass awareness at national level. Authentic literature and precise information on quinoa are not available to growers, researchers, educational institutions, industrialists and government departments engaged in reducing hunger and malnutrition problem. It is the main reason that Indian people are un-known with this valuable food crop and its nutritional as well as therapeutic benefits are still un-tapped. This manuscript, written in a comprehensive and lucid style, is intended to provide scientific, authentic and very useful information on various aspects of quinoa development in context to Indian conditions. The subject matter has been systematically arranged and is logical, precise and to the point.

I would like to express my heart-felt thanks and gratitude to Col. V. K. Sharma, Chairman, Indian Medicinal Plants Marketing Federation, Dr. A. K. Singh, DDG (AE), Indian Council of Agricultural Research, and Shri A. K. Singh, IAS, District Magistrate, Fatehpur, Uttar Pradesh, Dr. S. L. Jadhav, Director, Agriculture and Horticulture, Government of Maharashtra, Dr. J. P. Sharma, Joint Director, Indian Agricultural Research Institute and Dr. Dhoom Singh, Director (Agriculture Extension), Chandra Shekhar Azad University of Agriculture & Technology, Kanpur (U.P.) for their inspiration, encouragement and sustained support.

My heart-felt thanks are also due to Shri Dev Singh, Secretary General, Sarvati Devi Memorial Trust, Shri Dinesh Singh, Managing Director, 3D Industries Pvt. Ltd, Shri Mukesh Dave, President, Federal Union of Quinoa Growers and Traders of India and Shri Pradeep Dwivedi, Director, Nutrelis Agro Foods, Noida (U.P.) who extended ungrudging help, their co-operation and pivotal role in implementation of quinoa development programmes in India in collaboration with Indian Medicinal Plants Marketing Federation. I am much thankful to Hon'ble Shri Radha Mohan Singh, Union Minister of Agriculture and Farmers Welfare, Shri S.K. Pattanayak, IAS, Secretary, Ministry of Agriculture & Farmers Welfare, Govt. of India, Prof. Ashutosh Sharma, Secretary, Ministry of Science and Technology, Govt. of India, Shri J.P Meena, IAS, Secretary, Ministry of Food Processing Industries, Govt. of India, Shri Shyam Khadka, FAO Representative in India, Prof (Dr.) A.K. Srivastava, Chairman, Agricultural Scientists Recruitment Board (ICAR) and Hon'ble Vice-Chancellor, Agriculture University, Jodhpur (Rajasthan) for giving FOREWORD to this book. Further, I would like to express my acknowledgement and thanks to all authors whose papers I have consulted during preparation of this text.

I hope, this book would not only be more informative to its readers but also would promote a concern, that every citizen of India has an access to nutrient rich quality food for desired energy and healthy life.

The suggestions, comments and healthy criticisms to improve the book are heartly invited.

Dr. Dharm Singh

Contents

Chapter 1

FOOD AND NUTRITIONAL SECURITY – A MAJOR CHALLENGE

1.1 Introduction

Food is a basic human right, key to health and one of the requirements for the survival of human beings. A good and varied diet gives us energy, helps us to grow and enable us to fight off infections. Nutrients are the part of food that the body uses to build and repair tissuses, produce energy to fuel growth, movement and keep warm as well as protect from infections.

Hunger is detrimental to developing countries as it pushes improverished families into downward spiral and prevents further development. Eliminating hunger and malnutrition are one of the most fundamental challenges facing humanity throughout worldover. Malnutrition is one of the most devastating problems worldwide and is inextricably linked with poverty. Malnutrition and its associated diseases can be caused by eating too little, eating too much or eating an unbalanced diet that lacks necessary nutrients. The problems arise from cultural, political and economic realities that must be addressed in tandem. However, one significant step should be taken to educate and fight terrible problem of malnutrition. Deficiencies in nutrition inflict long-term damage to both individuals and society. Nutrition-deficient individuals are more likely to have infectious diseases, such as pneumonia and tuberculosis, which lead to a higher mortality rate. In addition, nutrition-deficient individuals are less productive at work. Low productivity not only gives them low pay that traps them in a vicious circle of under-nutrition but also brings inefficiency to the society, especially in India where labour is a major input factor for economic production.

On the other hand, over-nutrition also has severe consequences. In India national obesity rates is 14 % for women and 18 % for men with some urban areas having rates as high as 40 %. Obesity causes several non-communicable diseases such as cardio-vascular diseases, diabetes, cancers and chronic respiratory diseases.

- **Hunger:** Hunger is a term which has three meanings i.e., painful sensation / exhausted condition caused by want of food, scarcity of food in the country and a strong desire / craving. The Food and Agriculture Organization of the United Nations (FAO) defines food deprivation or under-nourishment, as the consumption of fewer than about 1800 kilo calories a day—the minimum that most people require to live a healthy and productive life. According to the author of present text, "Food should be provided to all as a matter of right without inflicting any humiliation on the poor." The hungry in the world are hungry because they are poor. They are poor because they own too little resources of land, capital or skills. Hunger is primarily a problem of poverty and not of food production.

- **Under-nutrition:** It goes beyond calories and signifies deficiencies in any or all of the following energy sources, protein, essential vitamins and minerals. Under-nutrition is the result of inadequate intake of food in terms of either quantity or quality, poor utilization of nutrients due to infections or other illnesses or a combination of these factors. These, in turn, are caused by a range of factors including household food insecurity; inadequate maternal health or child care practices or inadequate access to health services, safe water and sanitation. Under-nutrition contribute to nealy half of all deaths in children under 5 and is widespread in Asia and Africa. This translates into the unnecessary loss of about 3 million young lives a year. Under-nutrition puts children at greater risk of dying from common infections, increases the frequency and severity of such infections and contributes to delayed recovery. In addition, the interection between under-nutrition and infection can creat a potentially lethal cycle of worsening illness and deteriorating nutritional status.
- **Malnutrition:** Malnutrition refers more broadly to both under-nutrition (problems of deficiencies) and over-nutrition (problems of un-balanced over diets). It is the situation where there is an unbalanced diet in which some nutrients are in excess, lacking or in wrong proportion. Simply, we can categorise it to be under-nutrition and over-nutrition. This often happens as a result of the inability to access food because of lack of purchasing power which includes consuming too many calories in relation to energy requirements, with or without low intake of micro-nutrient rich foods.

  Malnutrition is basically of two types:

  - The first and most important is protein energy malnutrition (lack of calories and protein). It is more lethal form of malnutrition and leads to growth failure i.e. wasting (characterized by rapid weight loss due to lack of protein and in severe case cause death). Deficiency in macro-nutrients such as protein, carbohydrates and fat provoke Protein-Calorie Malnutrition. Protein-Calorie Malnutrition, also known as "protein-energy malnutrition", is defined by the WHO as being a pathological condition that results from a lower ingestion of protein and calories, which occurs more frequently in children under five years of age. Macro-nutrients when combined with micro-nutrient deficiencies, they are among the most important nutritional problems with millions of pregnant women, elderly and particularly, young children affected.
  - Second one is deficiency of nutrients (vitamins, micro-nutrients and minerals) caused stunting which is a slow and cumulative process. Iron, Vitamin-A and Iodine are three very important micro-nutrients responsible for malnutrition. Iron deficiency cause anemia in children and pregnant women. Iodine deficiency cause impaired congnitive development in children, spontaneous abortion and congenital abnormalities. Vitamin-A

deficiency cause night blindness, reduces resistance power of body to diseases and growth retardation in children.

Malnutritiion can also be caused by diseases, like diarrhea by reducing body's ability to convert food into usable nutrients. The establishment and severity of malnutrition depends on the cause, intensity and duration of the nutritional deficiency. It can be caused, primarily, by an inadequate diet or, secondarily, by deficiency in gastro-intestinal absorption and / or increase in demand, or even, by an excessive excretion of nutrients.

- **Poverty:** Poverty is socio-economic phenomenon whereby the source available to a society are used to satisfy the wants of a few, while many do not have even the basic needs. Absolute poverty means that the person does not have the required purchasing power to purchase the food necessary to provide the minimum nutritional intake of calories. Absolute poverty is a condition of life so characterised by malnutrition, illiteracy and disease as to be beneath any responsible difination of human decency.

1.2 Global Hunger Index (GHI)

It is a multi-dimentional statistical tool used to describe the state of countries hunger situation. It is released annually by US based International Food Policy Research Institure (IFPRI) since 2006. It ranks countries on a 100 point scale. Zero on the scale is the best score which indicates no hunger in the country and 100 is the worst. It highlights success and failures in reduction of hunger.

1.3 State of global hunger and malnutrition

Approximately 842 million people suffer from hunger worldwide that is almost 12 % of the world's population. About 98 % of those who suffer from hunger live in developing countries. The 2015 Global Hunger Index shows that the level of hunger in developing countries as a group has fallen by 27 %. Yet the state of hunger in the world remains serious. Levels of hunger are alarming or serious in 52 countries. India has highest population of hungry people in the world. Those worse than India are Niger, Chad, Ethiopia, Afghanistan and Pakistan. The most two important countries with alarming GHI scores are Africa and south of the Sahara. Brazil and Argentina ranked the best among developing nations.

According to projections from the Food and Agriculture Organization (FAO) of the United Nations, about 795 million people worldwide, roughly one in nine, are estimated to be chronically under-nourished in 2014–2016. Over 60 % of the world hungry are women. Almost all the hungry people (780 million) live in developing countries, representing 12.9 % or one in eight of the population of developing countries. Children are the most visible victims of under-nutrition. As of 2013, 161 million children, approximately one in four, were stunted and 51 million children suffered from wasting. Nearly half of all child deaths under age five are due to malnutrition, which claims the lives of about 3.1 million children per year.

Malnutrition is one of the most important underlying causes of child mortality in developing countries, particularly during the first 5 years of life, the major causes for this are poverty, world conflicts, lack of education, natural disasters and poor access to health care. PCM usually manifests early in children between 6 months and 2 years of age and is associated with early weaning, delayed introduction of complementary foods, a low-protein diet and severe or frequent infections.

Nearly one-third of children in the developing world are malnourished. Every second pregnant woman and about 40 % of pre-school children are estimated to be anemic in developing countries. An estimated 250 million pre-school children are vitamin-A deficient. About 2.5 to 5 lakh Vitamin-A deficient children become blind every year, half of them dying within 12 months losing their eye sight. Nutritional status affects every aspect of a child's health, including normal growth and development, physical activity and response to serious illness.

The real facts of world hunger are summarized below:

- 11.3 % of the world's population i.e., 805 million is hungry.
- 1 in every 9 people goes to bed hungry each night.
- 98 % of the world's hungry live in developing countries. Most of them live in Asia.
- 60 % of world's hungry are women and girls.
- Many hungry people live in countries with food surpluses, not food shortages.
- 75 % of world's poorest families don't buy their food, they grow it.
- One-third of the food produced around the world is never consumed.
- Hunger traps people into a life of poverty and ultimately, more hunger.
- Nearly 20 % of all children around the world are undernourished.
- Hunger kill more people each year than AIDS, malaria and tuberculosis combined.
- Every 10 seconds, a child dies from hunger.
- There will be over 2 billion more people who need food by 2050.
- The world produces enough food for everyone to live a healhy, productive life, but those who go hungry either do not have land to grow food or money to purchase it.
- Female farmers have the potential to pull 150 million people out of hunger.
- Poverty is the principal cause of hunger worldover.
- 17 million children are born from the mothers who not get proper nutrition during their pregnancy.

1.4 State of hunger and malnutrition in India

India is one of the fastest growing countries in terms of population and economic growth. Since Independence in 1947, its economic status has been classified as a low-income country where most of the population is still living below the National Poverty Line. In 2016, the International Food Policy Research Institute (IFPRI) placed India at 97 among 118 developing countries as per Global Hunger Index. Amongst South Asian nations, India ranks third behind only Afghanistan and Pakistan with a Global Hunger Index score of 29.0 (a serious situation). Neighbouring countries that scored better are Sri Lanka, Bangladesh, Nepal and China. India has 194.6 million hungry people that is one-fourth of the world's population.

Under-nutrition is more prevalent in rural areas, mainly due to low socio-economic status. Anaemia for both men and women is only slightly higher in rural areas than in urban areas. For example, in 2005, 40 % of women in rural areas and 36 % of women in urban areas were found to have mild anaemia. In urban areas, overweight status and obesity are over three times as high as rural areas. Malnutrition among women has long been recognized as a serious problem in India, but national-level data on levels and causes of malnutrition have been scarce.

The World Bank and FAO estimates that India is one of the highest ranking countries in the world for the number of children suffering from malnutrition. The prevalence of under-weight children in India is among the highest in the world and is nearly double that of Sub-Saharan Africa with dire consequences for mobility, mortality, productivity and economic growth. India is home to 40 % of the world's malnourished children and 35 % of the developing world's low-birth weight infants. Every year, 2.5 million children die in India, accounting for one in five deaths in the world. More than half of these deaths could be prevented if children were well nourished. Despite India's 50 % increase in GDP since 1991, more than one third of the world's malnourished children live in India. Among these, half of them under 3 are under-weight and a third of wealthiest children are over-nutriented. In India, 44 % of children under the age of 5 are under-weight. About 72 % of infants and 52 % of married women have anaemia. Studies have conclusively shown that malnutrition during pregnancy causes the child to have increased the risk of future diseases, physical retardation and reduced cognitive abilities.

In terms of geographical regions, Madhya Pradesh, Jharkhand and Bihar have very high rate of under-nutrition. States with lowest %age of under-nutrition include Mizoram, Sikkim, Manipur, Kerala, Punjab and Goa, although the rate is still considerably higher than that of developed nations. Further, anaemia is found in over 70 % of individuals in the states of Bihar, Chhattisgarh, Madhya Pradesh, Andhra Pradesh, Uttar Pradesh, Karnataka, Haryana and Jharkhand. Less than 50 % of individuals in Goa, Manipur, Mizoram and Kerala have anaemia. Punjab, Kerala, and Delhi also face the highest rate of overweight and obese individuals.

One of the major causes for malnutrition in India is gender inequality. The main reasons for bad situation of malnutrition in India are poverty, un-employment, lack of sanitation and safe drinking water, and lack of effective health care. Due to the low social status of Indian women, their diet often lacks in both quality and quantity. Women, who suffer malnutrition, are less likely to have healthy babies. In India, mothers generally lack proper knowledge in feeding children. Consequently, new born infants are unable to get adequate amount of nutrition from their mothers.

The key facts about hunger in India are given below:

- India is home to the largest malnourished population in the world.
- 194.6 million people go hungry every day, that is one-fourth of the world's population.

- 15.2 % of population is undernourished.
- 30.7 % of children under 5 years of age are under weight.
- 38.7 % children under 5 years of age are stunted.
- One in four children are malnourished.
- 3000 children die every day from poor diet related illness.
- 24 % children of under 5 years of age deaths every year.
- 30 % children are get neo-natal death each year.

1.5 Food security

Food security is a bigger issue beyond just providing a basic human need and right. It is an integral part of human development. Food security has four basic components; *availability, accessibility, utilization* and *stabilization.* It is essential to ensure availability of food grains to the common people at an affordable price. In the contest of persistently high food price inflation over the last few years, it was argued that the supply constraint causing high food prices was rooted in the slow growth of agriculture. Therefore, the challenge of meeting the food requirement of an ever increasing population can only be met by practicing sustainable agriculture, protecting natural resources and using production technologies that conserve and enhance the natural resource base of crops.

Key factors responsible for food crisis

A terrible situation of food crisis is emerging globally as a major threat to human survival. The key factors responsible for emerging situation of food crisis are given below:

- Diversion of agricultural land and fragmentation of holdings to non-agricultural purposes.
- Failure of traditional crops i.e., wheat, paddy, maize etc. to sustain agricultural production and profitability under marginal environment.
- Increase in water crisis, rainfed areas, droughts, soil salinity and climate change.
- Abandoning of farming occupation, ageing of cultivators and lack of interest of youth in agriculture sector.
- Major post-harvest losses resulting low availability of foodgrains for consumption.

Climate change is increasingly viewed as a current and future major cause of hunger and poverty. Increasing drought, flooding and changing climate pattern requiring a shift in crops and farming practices.

Key factors responsible for malnutrition

The key factors responsible for malnutrition worldover are as follows:

- Insufficient supply of protein, energy and micro-nutrients.
- Insufficient household food security.

- Unhealthy environment.
- Mal-functioning of society, injustice, war, natural disaster etc.
- Poor education of women.
- Insufficient child maternal care.
- Severe or frequent infections of diarrhea, pneumonia, tuberculosis etc.
- Poor utilization of nutrients due to infections or other illnesses or a combination of these factors.
- Lack of sanitation and safe drinking water.

1.6 Food loss and food wastgage

It is estimated that nearly one-third of the food produced in the world for human consumption every year gets lost or wasted. About 40 % of the fruits and vegetables, 30 % of food grains and 20 % animal meat are lost due to insufficient supply chain management and do not reach the consumer markets. While significant levels of food losses occur upstream at harvest and during post-harvest handling, a lot of food is lost or wasted during the distribution and consumption stages. Some food is also wasted on the shelves and in the warehouses of food businesses either due to excess production, introduction of new products, labelling errors or due to shorter remaining shelf-life. Such food could be salvaged by timely withdrawing it from the distribution net-work, aggregating it and then re-directing it to the people in need. On otherhand, 8 to 10 % food is wasted by the people during end of supply chain which incluses 56 % during wedding-bunquets, 21 % in Hotels, 17 % in restaurants and remaining 6 % in homes. This total annual loss of food in India costs Rs. 1 Lakh crore per year which is huge drain of precious money of India.

1.7 Need to solve problem of food insecurity and mal-nutrition in India

Green revolution in wheat, white revolution in milk production and IT revolution in India put on the path of prosperity. But, health care remains a missed opportunity alongside manufacturing, tourism and nuclear energy. Lack of adequate food or inadequate nutritive value in the food causes malnutrition. India runs two of the world's biggest Child Development programmes i.e., Integrated Child Development Scheme (ICDS) and Mid-day meal programme. But, both schemes have failed to reach out to the most vulnerable children and in reducing malnutrition in children considerably. India needs a lot more to be done and also needs another revolution to reduce malnutrition.

About 70 % of world's poor people live in rural areas and depend on agriculture and allied activities for their livelihood and 50 % hungry people are farm families. Over 75 % of the worlds poorest grow their own food. This causes widespread food insecurity in developing countries, as drought, climate change and natural disaster can easily cut off a family's food supply. Therefore, any revolution in Farm sector worldover could change the scenerio of malnutrition.

1.8 Measures to eradicate poverty and hunger

Following are the suggestions to eradicate poverty and hunger in India:

- Ensure political, social and economic environment for the eradication of poverty and for equal participation of men and women to achieve sustainable food security to all.
- Ensure food and agricultual trade over all trade policies or businesses to foster food security for all through a fair and market oriented world trade system.
- Promote proper alloction and use of public and private investments to foster sustainable agricultural system in high and low political area.
- National and local governments in the states have a prime responsibility to ensure food security.
- The participation of people's organizations and NGOs at all levels must be strenghtened and deepened.

At the national level, following political issues must be taken:

- Economic decentralization and self-governance in agriculture should be made as a part of the responsibility of the panchayat.
- Maintenance of biodiversity, diversity in agriculture, increased access in credit and other resources, self-reliance and self-sufficiency should be made at the panchayat level. Promote a perspective which is based on local food security.
- Streamline and strengthen the PDS, and bring them under the control of Panchayt so that the local people can establish their control and decision over their right to food.
- The NGOs in the country should play a role of advocacy and awareness building. With a national and international will power, the NGO's role and with people's participation, food security can be fully attained whereby poverty and hunger can be eradicated.

1.9 Role of Quinoa in food security

Quinoa crop could be the silver bullet that addresses both food security and health security in the next century. The main reasons of its selection as food for future are its high nutritional value, well-balanced amino acid content in the grains, wide adaptability to climate change and harsh environments, drought, salinity and its power as staple food of world's poorest people. It has the highest potential of next revolution, solution of malnutrition and other health problems. While doing so, quinoa can help small farmers and boost commercial exports. Its specialities, nutritive values, properties, benefits etc have been discussed in different chapters of this text.

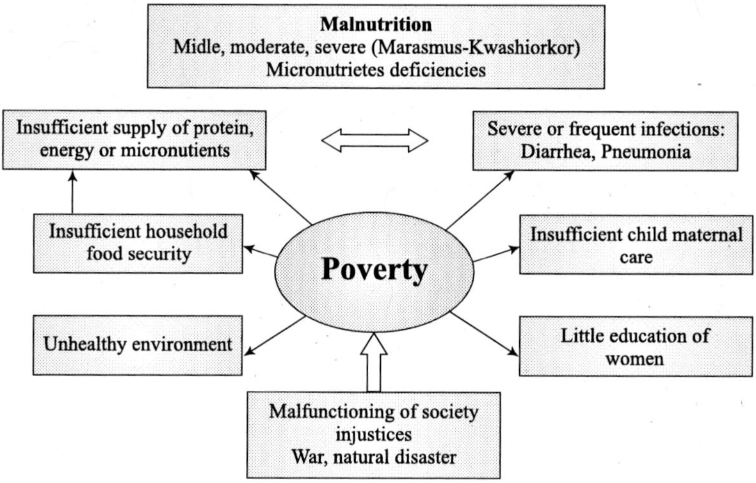

Fig. 1. *Key factors responsible for malnutrition*

Chapter 2

The Quinoa – A Healthy Food of Future

2.1 INTRODUCTION

Majority of the plant-based foods available in the markets worldover are unable to fulfil our nutritional requirements. Due to this fact, we need to combine two or more foods in order to make nutrient-rich diet. It has found necessary to combine nutritionally valued crops i.e., soybean and wheat to obtain a suitable amino acid pattern to meet nutritional needs. Although, animal-based foods like meat, dairy products and eggs generally provide a complete protein, but obviously these are not options for vegans and vegetarians of India. Therefore, now a days, new food products featuring ancient grains are appearing and attracting attention worldwide because of their high nutritional value, high quality protein and potential of wide adoptability under different climatic conditions. Among such foods, quinoa is one of the most important nutrient rich foods gaining fast popularity worldover.

Quinoa is a grain crop grown primarily for its edible grains to support nutritional requirements. It provides a complete protein, means a source of all nine essential amino acids in the right proportions. The high Lysine value in quinoa make it of special interest, which is unusual in the plant kingdom. Quinoa is also rich in the essential sulfur containing amino acids i.e., methionine and cystine. Amino acid pattern of quinoa is very similar to the combination of soybeans and wheat and may provide a suitable alternative. While no single food can supply all the essential life sustaining nutrients. Its consumption increases power and stamina of the people. Therefore, quinoa has been termed as "Super-food", "Mothers of all grains", "Food for future", "Power food", "Gold of people"and "Food for global health security".

Food and Agriculture Organisation (FAO) has considered Quinoa as healthy Food and identified it as one of the crops to provide food security in the next century. The main reasons of its selection as food for future are its high nutritional value, well-balanced amino acid content in the grains, wide adaptability to climate change and harsh environment, drought, salinity and its power as staple food of world's poorest people. It is classified by FAO in 1996 as one of the most promising crops of human being, not only due to beneficial properties and its diverse uses, but also considering it as an alternative to solve the serious problems caused due to nutritional deficiencies. NASA also considered quinoa as best food within "Controlled Ecological Life Support System" to equip its rockets in long-duration space travel and to solve the problems of inadequate protein intake.

India needs several million tonnes of quinoa a year to meet the nutritional requirement of poor people, diabetics, cardiac patients and malnourished children.

Quinoa can be another cash crop for millions of farmers and raw material for food industries. While doing so, it can help small farmers and boost commercial exports.

2.2 SPECIAL FACTS AND FEATURES OF QUINOA

Quinoa has several specialities over common food grains, thereby, attracting people worldover. Its general, genetical, nutritional and therapeutic specialities alongwith important facts have been summarized here under.

General facts and features:

- Botanically, quinoa not belongs to the family of grasses and cereals, but, it is pseudo-cereal because of its high starch content.
- Quinoa is not a grain at all. It is a seed, but, used as a grain.
- Quinoa is only the crop in the world with highest potential to provide food security, nutritional security, health security and better opportunities for the development of food industries.
- Quinoa is only the single food which can supply complete protein, all essential life sustaining nutrients and can reduce the risk of various diseases in very effective and preventive way.
- Quinoa plant is slow growing in initial stage, but, it grows very quickly once it reaches about one foot in height.
- The leaves of quinoa are coloured due to the presence of betacyanins.
- Quinoa seeds do not exhibit dormancy. Seeds may remain in the soil for two or three years without germination and germinate immediately when available conditions are suitable.
- Quinoa is more preferred than other whole grain like rice as it takes very less time (10 to 15 minutes) in cooking as compared to other whole grains. Cooked grains of quinoa become larger in size atleast 4 times.
- Quinoa is considered as whole grain food and it is slowly digested as compared to other refined grains. As a result, it feeds for a longer period of time with more stabilization of blood sugar preventing fluctuation in energy level.
- The stems are hollow, spongy inside and rich in pectin and cellulose that can be used for the manufacturing of paper and paper board.
- Essential amino acids are located in the nucleus of the grain, while in other cereals, like rice or wheat, they are present in their exosperm or hull.
- Highest protein content, excellent taste of cooked grains and high content of quality protein makes quinoa perfect food for vegetarian and vegans.
- Quinoa leaves have granules on surface which contain cells rich in calcium oxalate. These leaves are able to retain a water film, which increases the relative humidity of the atmosphere surrounding the leaf thereby reducing transpiration.
- Quinoa cannot be used alone for bread making as its grains do not contain gluten (binding substance).

- With a similar caloric content to that of cereals, it is highly recommended food for people who want to loose weight in healthy and pleasurable manner.
- Quinoa is only the single food which can supply all the essential life sustaining nutrients for longer duration and can reduce the risk of various diseases in the preventive way. Therefore, technical report of NASA named quinoa during 2013 as great food to take it into the space.
- The root system develops from a tap root to form a highly branched system that makes plants more resistant to drought.
- Quinoa is highly resistant and adaptable crop in the world to wide range of climate change, salinity and drought with least effect by other climate change induced weather aberration, resulting no risk of crop failure under water stress / rain fed conditions.
- European quinoa is sweet which does not contain saponin on grains and not require the processing for removal of saponin. The outer layer contains minerals and dietary fibres which add nutritional value over saponin containing varieties.

Genetic features:

Quinoa has broad genetic variability. Its gene pool is extremely strategic for developing superior varieties of more grain size, excellent grain colour and resistance to biotic and abiotic factors, good grain yield and by-products. Study revealed that gene clusters involved in Abscisic Acid (ABA) signaling were substantially expanded in quinoa genome compared with other plant species in its family.

Nutritional feature:

- Quinoa nutritionally termed as super food.
- It provides gluten free and low-fat protein which make it an ideal food for the people suffering from celiac disease.
- Quinoa protein makes it an ideal food for combining in vegetarian diets where protein intake may be deficient for athletes.
- Quinoa grains contain all nine essential amino acids in the right proportions and greater amount of lysine and isoleucine making it a complete protein source.
- Quinoa grains are rich in nutrients and best source of dietary fibre, minerals and vitamins.
- Quinoa contains significant amount of certain tocopherols (group of Vitamin-E) largely absent in most food grains.
- It is a good source of various nutrients like folate, zinc and phosphorus in contrast to whole wheat, which does not qualify as a good source in our rating system.
- Quinoa contains highly basifying minerals like calcium and magnesium among its amino-acids which help in preventing and treating cardio-vascular diseases, very common in people with over-weight and obesity.
- Quinoa has more fibre than other cereals which helps in removing cholesterol.

Therapeutic features:

- Quinoa plays an important role as aphrodisiac in the problem of male potency i.e., problem of erectile dysfunction, very low seminal fluid and sperms, lack of sexual excitement and sexual desire.
- Lignant, a phyto-nutrient, is abundant in quinoa grains which is valuable nutrient for maintaining better health of our heart.
- Quinoa helps in reducing the LDL (bad cholesterol) in the body and increases HDL (good cholesterol) due to its omega-3 and omega-6 content.
- Quinoa has a low glycemic index making it an ideal food for people suffering from diabetes.
- Quercetin present in quinoa is beneficial in type-2 diabetes, asthma, inflammation, viral infections and in prevention of cancer and chronic prostrate infections. Quercetin is also beneficial in atherosclerosis (hardening of arteries), circulatory problems and heart diseases.
- Quinoa is fibre rich, cholesterol free and low fat food grain which reduces the risk of high blood pressure, heart attack, blood sugar level and helps in weight loss and obesity.
- Quinoa is loaded with anti-oxidants which neutralizes free radicals and are believed to help fight ageing and many diseases.
- Saponins of quinoa help in lowering blood cholesterol, inhibits growth of cancer cells, eliminate digestive toxins and strengthen immune system of human beings.
- Quinoa feeds for a longer period of time and prevent fluctuation in energy level with more stabilization of blood sugar.
- Quinoa helps in preventing and treating cardio-vascular diseases (common in case of over-weight and obesity) as it has high content of basifying minerals like calcium and magnesium.
- The magnesium and vitamin B-2 in quinoa help to relax blood vessels and improve the energy metabolism, thus, reduces the frequency of headache episode.
- Quinoa is a good source of dietary fibre needed for health blood sugar.
- Phenolic derivatives act as natural agents which have been proved to be very good anti-oxidants, scavenging free radicles and providing metal chelating activities.
- Poly-phenols of quinoa have been implicated in health benefits, such as prevention of cancer and cardio-vascular diseases.
- Quinoa is functional food that lowers risk of several diseases by protecting cell membrane with proven results in neurological functions.
- Quinoa possesses anti-gastric and demulecent properties.
- Quinoa has a more satiating power and is more effective for weight control and obesity than other cereals.

2.3 ADAPTABILITY TO CLIMATE CHANGE AND MARGINAL ENVIRONMENT

Quinoa has wide range of remarkable adaptability to different agro-ecological zones with high tolerance to drought, frost and infertility. It can easily withstand severe frost exposure, low radiation stress, drought and salinity and is considered one of the highly adaptable crops in the world for its growth and yield in a wide range of climate and environments. National Institute of Agriculture and Forestry, Bolivia, ranked quinoa as most resistant crop to climate change.

Quinoa is largely grown in the Andean Region and the areas of Altiplano falls in Peru and Bolivia which has one of the most difficult ecologies for modern agriculture. Quinoa survives in this ecological environment of low biotic interactions. Altiplano altitudinal limits are 3000 to 4000 metres above sea level, where soils are often alluvial and poorly drained. Rainfall varies greatly among the different areas of cultivation in South America. In the Ecuadorian Andean, it is 600 to 880 mm, in the Mantaro Valley, it is 400 to 500 mm and in the Lake Titicaca area, it is 500 to 800 mm. However, as one moves south of the Altiplano of Bolivia and Northern Chile, precipitation decreases to levels of 50 to 100 mm, conditions in which quinoa is also produced. Bolivia's Southern Altiplano is considered the main geographic area where the quinoa is grown to feed the international demand. On the other hand, quinoa is also produced between the eighth and ninth region of Chile with rainfall above 2000 mm and considering the conditions where quinoa is cultivated and the genetic variability available.

2.4 ROLE OF QUINOA IN HUMAN NUTRITION AND HEALTH

The emerging situation of food crisis world over has reinvigorated the efforts to search an alternative foods for future generation for their survival. FAO identified only two staple foods for the growing population i.e., quinoa and edible insects. Being as a vegetarian country, Indian people cannot accept eating of edible insects as food. Quinoa can be considered future vegetarian food as it is only the highly potential crop to provide food security in the next century. As per FAO, quinoa attained top position to provide food security, health security, livelihood security and better opportunities for the development of food industries. It is versatile and can substitute any food-grain in any recipe. Its advantages over traditional food crops / cereals are given hereunder:

- Better adaptability to harsh environments of the wide range of areas.
- The nutritional well-balanced amino acids content of the grain.
- Its power as food, being a staple in the diet of world's poorest people.

In comparision to cereal crops like wheat, quinoa is higher in fat content and can provide valuable amount of heart-healthy fats like mono-unsaturated fat (in the form of Oleic acid). Quinoa also provide small amount of the Omega-3 fatty acid and alpha-linoleic acids (ALA). Quinoa does not get oxidized as per recent studies.

Quinoa is considered a nutritious food and can be a great part of a balance diet. It was important to the diet of pre-Columbian Andean civilizations. Protein content is very high (14 % by mass) in quinoa as compared to other cereal/pseudo-cereals. The calories content per 100 gm is higher than brown rice, potatoes, barley and millet, but is less than wild rice and oats. Nutritional evaluations indicate that quinoa is a source of complete protein which is relatively high in essential amino acids. Compared to other plant foods, quinoa is generally higher in protein than most grains. Quinoa also has a favourable balance between its essential amino acid content compared to other plant foods. Finally, quinoa is a good source of energy and dietary fibre and has significant amounts of minerals such as iron and zinc.

Table 1: Nutrient contents of quinoa and other food grains
(Per 100 gm dry weight)

Constituents	Quinoa	Maize	Rice	Wheat
Energy (Kcal)	399	408	372	392
Protein (g)	16.5	10.2	7.6	14.3
Fat (g)	6.3	4.7	2.2	2.3
Total carbohydrate (g)	69	81.1	80.4	78.4
Iron (mg)	13.2	2.1	0.7	3.8
Zinc (mg)	4.4	2.9	0.6	4.7

Quinoa is a rich source of the B-Vitamins – thiamine, riboflavin, niacin and pantothenic acid, vitamin B-6, vitamin-E and folate, and is a good source of the dietary minerals i.e., iron, magnesium, phosphorus, potassium and zinc. The pseudo-cereal contains a modest amount of calcium and thus is useful for vegans and those who are lactose intolerant. It is gluten-free and considered easy to digest. Due to these characteristics, it is being considered a possible crop in NASA's Controlled Ecological Life Support System for long-duration human occupied space flights. The grain may be germinated in its raw form to boost its nutritional value, provided that the grains are rinsed thoroughly to remove any saponin. It has a notably short germination period, only 2–4 hours in a glass of clean water is enough to make it sprout and release gases, as opposed to 12 hours with wheat. This process, besides its nutritional enhancements, softens the seeds, making them suitable to be added to salads and other cold foods. Other pseudo-grains are similar in complete protein levels; protein in buckwheat is 18 % compared to 14 % in Quinoa; Amaranth, a related species to Quinoa, ranges from 12 to 17.5 %.

With the emerging health consciousness, the consumer's trend towards ancient grains is increasing day by day. Quinoa is available in big Bazars throughout world and recently gained worldwide importance due to its nutritional benefits. The nutritional value of its grains has been widely recognized for their high quality protein (particularly rich in essential amino acids) and for their carbohydrate, oil, mineral and vitamin content. Quinoa is also considered a good source of dietary fibre and other bioactive compounds, such as poly-phenols and tri-terpenoids.

In this context, various studies adopting different biological approaches have confirmed that the bioactive components present in both seeds and leaves of quinoa possess hypo-cholesterolemic, anti-oxidant, anti-inflammatory and anti-cancer effects, as well as safe for consumption. Quinoa, therefore, offers great potential for use in complementary and alternative medicine. With the emergence of new technologies in the area of chemical research, molecular biology and pharmacology, the use of quinoa as a nutraceutical agent is increasingly gaining recognition. Nevertheless, further work is needed to gain a better understanding of the pharmacological potential of these phytochemicals, including their metabolic profile, toxicity, interaction with other compounds, stability of formulations and dosage regime, as well as potential polymorphisms which could affect therapeutic efficacy. In short, the information supports the potential of quinoa as a food supplement that could enrich the normal diet as a source of functional compounds important for reducing chronic disease risk factors and opening new perspectives for using quinoa in bio-medicine.

2.5 FUTURE PROSPECTS OF QUINOA DEVELOPMENT IN INDIA

India, located between 8° to 38° N and 68° to 93.5° E, exhibits enormous diversity for agro-climatic regions and edapho-climatic conditions. An increasing population in this region of the world demands not only an increase in food grain production but also a shift towards environmentally sound sustainable agriculture. Quinoa can play a major role in future diversification of agricultural systems in India, not only in the Himalayan region, but also in the North-Indian Plains. Quinoa can be termed 'under-utilized', especially for India, since inspite of its wide adaptability, rusticity and nutritional superiority, its commercial potential has remained untapped. In India, a large portion of the population has little access to protein-rich diet, since rice and wheat are the principal food crops. Protein rich grains of quinoa can help to make diets more balanced in this region and can play an important role in combating'silent hunger' among poor populations in India who have little access to protein-rich diet. Besides this, improved technologies and link with other areas, like product development and marketing, would enable the industry to tap its potential for diverse application.

Quinoa is a new crop for India which has been overlooked by researchers, industrialists and commercial and experimental farmers till date. The results of Front-line demonstrations of quinoa organised by author of this book in the different areas i.e., Rajasthan (2003–04), Haryana (2005–06), Uttar Pradesh (2005–06), Maharasthra (2006–07) and Gujarat (2007-08) states revealed that quinoa can be grown commercially with great success throughout India as it responded well to varied agro-ecological situtations.

It is also proved that quinoa performed well in hydroponic culture in controlled environment and fast growth was recorded with good yield of grains in Haryana state. Given these encouraging observations, a project entitled "Evaluation of production potential of quinoa in marginal areas of Indian sub-continents" was planned by the author of this text and started its implementation on 2nd October, 2009 in 75 districts of 10 states in India. The 5 years studies (2009–10 to 2013–14) under this project revealed the great success of quinoa production and open the way of its commercial production in India.

2.6 CHALLENGES IN SCALING UP OF QUINOA PRODUCTION IN INDIA

Despite the increasing importance of quinoa, its high potential of wide adaptability, positive research outcomes in pilot studies and global recognition, there are still many constraints and issues to be addressed before quinoa becomes a crop of choice in marginal areas where other major crops have so far been dominant, but are now failing due to climate change and degradation of soil and water resources.

The Key challenges in quinoa development in India are given below:

- Limited availability of genetically pure seeds for cultivation outside its indigenous environment.
- Limited authentic knowledge of the best agronomic and management practices including nutrient and water requirements, pest and disease control, harvesting and processing under marginal growing conditions.
- Little awareness about quinoa's nutritional benefits and the intricacy to incorporate it into local diets in different regions.
- Lack of suitable marketing channels where the farming communities could sell their produce.
- Lack of processing and saponin removal facilities in India in different states.
- Lack of attention of Government departments , State Agriculture Universities, ICAR Research Institutions and other research Agencies towards taking planned research and extention programmes.

2.7 ROLE OF FAO AND ICBA IN QUINOA DEVELOPMENT IN THE WORLD

Recognizing quinoa's potential for marginal environments, ICBA has been leading a global quinoa program since 2007 in collaboration with national, regional and international research, government and donor organizations in the NENA region as well as Central Asia. The Food and Agriculture Organization (FAO) of the United Nations, specifically served as the Secretariat of the International Year of Quinoa. Since then, FAO has conducted activities around the world and particularly in the Near East and North Africa (NENA) region. In the NENA, FAO has implemented a project in eight countries with positive outcomes that have indeed shown quinoa as a potential alternative and profitable crop for the region.

2.8 INTERNATIONAL YEAR OF QUINOA – 2013

On 20 February, 2013, the 67th UN General Assebly declared 2013 as "International Year of Quinoa (IYQ)"during its plenary at UN Headquaters in New York, fulfilling the General Assembly Resolution A/RES/66/221 of 22 December, 2011.

This declaration was in recognition of the indigenous peoples of the Andeas, who have maintained, controlled, protected and preserved quinoa as a food for present and future generations. The primary aim of declaring IYQ was to utilize quinoa's biodiversity and nutritional value for food security and eradication of poverty, as the UN General Assembly also pointed out quinoa's nutritional qualities and its adaptability to different agro-ecological conditions. The FAO Director General Jose Graziano da Silva, called it an "ally in the fight against hunger and food insecurity" at the official launch of the International year of quinoa.

To mark the International Year of Quinoa (IYQ), 2013, the Central Reserve Bank of Peru launhed a limited series of one Nuevo Sol coins known as Recursos Naturales del Peru (Natural Resources of Peru). With remark that the new coins will be used nation wide feature, three of Peru's main natural resources: anchovies, cacao and quinoa highlighting the importance of this Andean grain. Closing ceremonies were held in Bolivia, Peru, Argentina and Columbia followed by New York. The FAO officially launched a book "Quinoa in the kitchen" to continue to promote awareness about the potential of this super food. On the World Food Day, 2013, quinoa was served in UN restaurants in Geneva which includes the palais des Nations, WIPO, UNHCR, INICEF and International Environment House.

The IYQ constitute the first step in an on-going process to focus world attention on the role of quinoa in providing food nutritional security and poverty eradication in support of the achievement of the internationally agreed millennium development goal. The IYQ is expected to be a catalyst to enable the exchange of information and to start to generate medium and long-term programmes / projects for sustainable development of the cultivation of quinoa nationally and globally. Ultimately, it would raise awareness of the importance of the production and consumption of quinoa and its contribution to food and health security. In the context of IYQ implementation, a master plan has been designed as a dynamic tool including three components i.e., information and communication, research-technology and marketing and fund raising and cooperation. The budget of master plan is US dollar 28,82,630 and it will be implemented locally, nationally and internationally.

Chapter

3

ORIGIN, HISTORY AND DISTRIBUTION

Quinoa (*Chenopodium quinoa Willd.*) is a domesticated pseudo-cereal and one of the oldest nutritionally rich crops of South-America. *Chenopodium* species *(Chenopodium quinoa and Chenopodium album)* have been cultivated for centuries as a leafy vegetable as well as an important subsidiary grain crops for human consumption and animal food stuff. The quinoa grains have high-protein content with abundance of essential amino acids and a wide range of vitamins and minerals. It is one of the best super food due to high-protein and a balanced amino-acid spectrum with high lysine (5.1–6.4%) and methionine (0.4–1.0%) contents.

Quinoa is the only plant food that contains all the essential amino acids, trace elements and vitamins without containing gluten, a harmful compound. Essential amino acids are found in the nucleus of the grain, while in other cereals, like rice or wheat, in which they are located in their exosperm or hull. Further more, the crop has a remarkable adaptability to different agro-ecological regions. It can be grown successfully in different agro-climatic regions due to its wide range of adaptability and tolerance to drought and infertility.

3.1 ORIGIN

The Andean regions of South America, the cradle of great civilizations such as the Inca and Tiahuanaco, is considered the center of origin of quinoa. According to scientific research, quinoa is originated from the shores of Lake Titicaca in Peru and Bolivia, "the highest most saline water reservoir in the world". Shores of Lake Titicaca are considered as the area of its greatest diversity and genetic variation. According to Vavilov, the Andean Region corresponds to one of the great centres of origin of cultivated species and within it, there are different sub-centres. Because of nutritional value, indigenous people and researchers call it "The Golden grain" of the Andes and "the mother grain" by the Incas.

For 7000 years, indigenous people have maintained, controlled, protected and preserved quinoa in different ecological zones, maintaining many different varieties in their natural state, through natural gene banks, based on the principles of complementarity, re-distribution and living in harmony with nature and Mother earth. It was the main food of the ancient cultures of the Andean and is distributed in different agro-ecological zones within the Region. Most of the quinoa in the world comes from the Andean regions of Peru and Bolivia. There are a few different varieties of quinoa like black, red and white, but the most common type is white quinoa. Sometimes, quinoa is sold as a blend of white, red and black, described as rainbow quinoa or tri-colour quinoa.

3.2 HISTORICAL, ARCHEOLOGICAL AND CULTURAL BACKGROUND

- **Historical background:** Quinoa is a crop with very special history. On the Peruvian-Bolavian alpine plain, it has been grown for ages together with corn and beans as a staple food of ancient Incans. Quinoa, in language of Incans, means "Mother of cereals". Quinoa was domesticated probably in area of Altiplano surrounding Lake Titicaca. Incans have highly developed agriculture, but Spanish conquerors pushed their traditions into the background and actually destroyed their clever agricultural system. New high yielding Europian cereals, like wheat and barley, subsequently replaced quinoa. There are some 150 years old records about quinoa in the Czech Republic concerning cultivation of this plant for ornamental purposes in botanical gardens. High nutritional value of quinoa, known from the beginning of the 20th century, caused a big boom during First World War in whole Europe including Czech Lands. Neverthless after some short time, the lack of growing experience left the crop behind again.

  Quinoa was cultivated and used by pre-Columbian civilizations and was replaced by cereals on the arrival of the Spanish, despite being a local staple food at that time. Existing historical evidence indicates that its domestication by the peoples of America may have occurred between 3,000 and 5,000 years BC. At the time of Spanish arrival, quinoa was well developed technologically and was widely distributed within and beyond Inca territory. The first Spaniard to note the cultivation of quinoa was Pedro de Valdivia who, on noticing the planted crops around Concepción, recorded that, for food, the native Indians also sowed quinoa among other plants. In his royal commentaries, famous writer Garcilaso de la Vega, describes quinoa as one of the second grains cultivated on the face of the earth, somewhat resembling millet or short-grain rice. He also mentioned the first shipment of seeds to Europe which were unfortunately dead on arrival and unable to germinate, perhaps because of the high humidity of the sea voyage. Later, it was reported that quinoa was cultivated in the highlands of Pasto and Quito, mentioning that little maize, but an abundance of quinoa was grown on these cold lands.

  The history of quinoa has little archaeological, linguistic and ethnographic evidence, as many religious rites associated with the use of the grain are not known. Evidence of the use of quinoa in Tiahuanaco pottery culture, represents the quinoa plant with several panicles distributed along the stem, which would show one of the most primitive races. Quinoa (*Chenopodium quinoa Willd*) is a domesticated species and cultivated in Peru since pre-Hispanic times (more than five thousand years). The area around Lake Titicaca is considered as the main centre of origin of quinoa and conservation centre for the greatest biological diversity of this species, in which there are ingenious farming systems and food culture that incorporates the daily digestion of a variety of grains. During the empire of the Incas, quinoa became one of its main crops and staple diet of the inhabitants. In late 70s, *Chenopodium quinoa* became an interesting crop due to increased demand of healthy nutritious food.

First more important trials were performed in England, Netherland, Germany, Denmark and Finland.

Despite being a fully domesticated species, the grains still contain saponin, so its removal is necessary before any type of consumption. Marginalization and replacement of this crop began with the Spanish conquest and the introduction of cereals such as barley and wheat. In this regard, it is indicated that the crop was never lost among the Andean people, but went un-noticed among the urban population of the region for mainly economic and social reasons. The economic crisis in the Andean countries during the 80's, established new patterns of economic development which differed from traditional ones, as they contemplated the development of non-traditional export systems, taking into account the opening of new markets in Europe and the United States, especially for food products such as quinoa. The welfare achieved by developed countries has led to the expansion of their food consumer markets in the search for new foods, often linked to ancient cultures. This situation has meant that quinoa has been transformed from a subsistence crop into a product with export potential.

- **Archaeological background:** There are archeological discoveries of quinoa in tombs of Tarapacá, Calama and Arica, in Chile and in different regions of Peru. Quinoa seeds were found in indigenous graves of Tarápaca, Calama, Tiltel and Quillagua, demonstrating that this crop has been cultivated since ancient times. Archeological findings in northern Chile have shown that quinoa was used prior to 3000 BC. In Ayacucho, Peru, evidence has been obtained that quinoa was cultivated there before 5000 BC. The quinoa plant was widely cultivated in the whole Andean region, in Columbia, Equador, Peru, Bolivia and Chile before the Spanish conquest. However, the habits and traditional foods of natives were replaced with foreign crops, such as wheat and barley. Therefore, quinoa was cultivated either in small plantations in rural areas for domestic consumption or as borders for other crops such as potatoes or maize. For this reason, it was classified as food for poor people.

  Archaeological evidence in the region of Ayacucho, Peru indicates that the domestication of quinoa started 5,000 years before Christ. There are also archaeological findings of quinoa consisting of seeds in ancient tombs of Tarapaca, Calama, and Arica and in other regions of Peru. Upon the arrival of the Spaniards, quinoa already had an advanced technological development and a wide distribution within the Inca territory and beyond. The first finding and report of quinoa from an outsider was made by Spaniard Pedro de Valdivia who was observing the cultures around Concepción, mentioned that the natives also grow quinoa among other plants for the diets. Subsequently, Bernabé Cobo confused Quinoa with Amaranth and indicates that quinoa is very similar to that plant in Europe.

  Quinoa is one of the oldest crops in the Andean Region, with approximately 7000 years of cultivation and great cultures, such as the Incas and Tiahuanacu have participated in its domestication and conservation. Quinoa was widely cultivated in the Andean Region by pre-Columbian cultures and its grains have been used in the diet of inhabitants of the valleys and in drier areas, having less than 350 mm of precipitation with higher altitudes (above 3500 m) and colder temperatures (average 12°C) such as the Altiplano.

- **Cultural background:** It argues that quinoa replaced corn in the entire Andean Region from *Colombia* to the south of Chiloe Island, having also acclimatized in many parts of the coast. It is believed that in Colombia, quinoa was introduced too late to address the lack of corn in some cold area and its undoubtable that during the Inca Empire, quinoa was wide spread in Ecuador, Peru, Bolivia, Chile and north-west Argentina, given its economic importance. Since apart from their protein rich seeds, its leaves were also used in salads. However, genetic erosion has been intense since the discovery of America. Most of the botanist mentioned that in the 16th century, there were more species domesticated in the Andean region than in Asia or Africa, mainly due to the replacement of local crops by others brought from Europe. This erosion was emphasized by social changes and people's attitudes about foreign crops, where a social prestige was created which led to Andean crops being under-valued. The survival of Andean crops was made possible by the peasant communities that inhabit the area, who based on their traditions and ancestral knowledge about the management and use of these species have managed to prevent their loss.

3.3 GEOGRAPHIC DISTRIBUTION

The geographical distribution of quinoa is mainly confined to the Andean region which extends from 5^0 North latitude to 43^0 South latitude including Colombia, Ecuador, Peru, Bolivia, Argentina and Chile. The highest level of quinoa production takes place in Bolivia, Peru and Ecuador. Its altitudinal distribution ranges from sea level to 4000 meters above sea level (masl), where the biggest genetic diversity is found in the Altiplano (high plain Lake Titicaca basin) regions of Bolivia and Peru. Quinoa is distributed throughout the entire Andean Region, from Colombia (Pasto) to northern Argentina (Jujuy and Salta) and Chile (Antofagasta), where a group of quinoas have been found at sea level at the Bío Bío Region. While most quinoa is still grown in Western South America from Columbia to Chile, it is also cultivated in the USA (Colorado and California), China, Europe, Canada and India. It is also cultivated experimentally in Finland and the UK. Increasing amounts are being exported to the developed world like Europe and the USA. In Europe, quinoa was introduced in England in the 1970s and later research projects focused on its production for humans and/or as a fodder crop under temperate conditions. The geographical distribution of quinoa in the Region extends from latitude 5 degrees North in Southern Colombia, to latitude 43° South in the Xth Region of Chile and its altitudinal distribution ranges from sea level in Chile to 4000 m in the Altiplano of Peru and Bolivia. Thus, existing quinoa of the coast valleys, inter andean valleys, Puna and Altiplano. Currently, quinoa is in a process of expansion because it has great potential to improve the living conditions of people in the Andean and the modern world.

The distribution starts from Narino to the Salares of southern Bolivia that includes countries like Ecuador, Peru and northern Argentina (Jujuy and Salta provinces).

The Atacama Desert forms a break in the distribution of the crop, which continues further south into Chile. Recently, it has been introduced in Europe, North America, Asia and Africa. Many European countries are members in the project entitled 'Quinoa—A multi-purpose crop for European Countries agricultural diversification' which was approved in 1993. The American and European Test of quinoa have yielded good results and demonstrate the potential of quinoa as a grain and fodder crop.

Today, Quinoa is grown in Peru, Bolivia, Ecuador, Chile, northern Argentina and other countries. Peru and Bolivia are the largest producers of this Andean grain that is similar to rice in colonial times, where the Spaniards called it "American arrocillo" or "wheat of the Incas". Peru possesses a genetic diversity of both wild and cultivated quinoa, making it one of the largest producers and exporters, which represent a large potential business opportunity for Peruvians as a whole.

The geographic distribution of world quinoa production is shown in Figure-2.

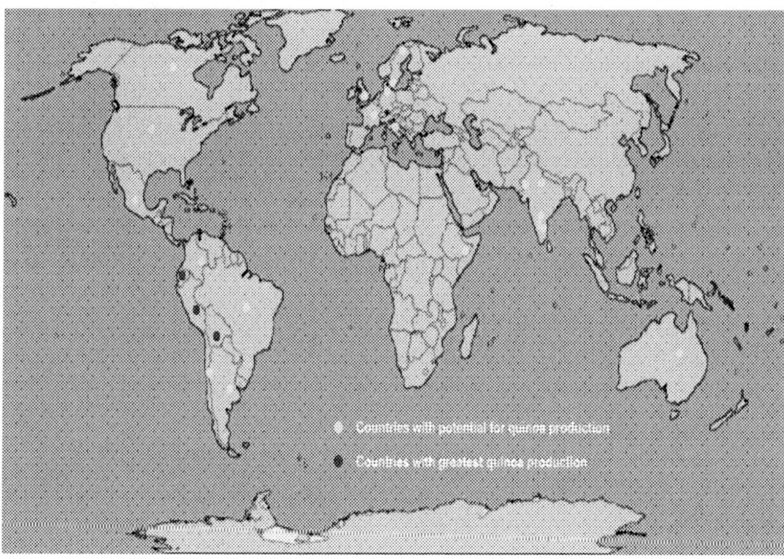

Fig. 2: *Quinoa producing countries*

In quinoa distrtibution map, it can be seen that it is distributed in countries like Bolivia, Peru, Ecuador, Europe, Asia, Africa, Australia, North America and India.

3.4 GLOBAL PRODUCTION SCENERIO

- **Quinoa producing countries:** Peru, Bolivia and Ecuador are the the largest producers of Quinoa in the Andean Region and the world. Until 2008, the production of quinoa in Peru and Bolivia accounted for 92 percent of quinoa produced in the world, while contribution of Ecuador was 4 percent. Quinoa production has increased in the last 20 years, especially in Bolivia.

Table 2: *Top Quinoa Producing countries of the World (thousand metric tons)*

Countries	Years								
	1961	**1970**	**1980**	**1990**	**2000**	**2010**	**2011**	**2013**	**2014**
Peru	22.5	7.3	16.3	6.3	28.2	41.1	41.2	52.2	114.3
Bolivia	9.2	9.7	8.9	16.1	23.8	36.1	38.3	50.5	77.4
Ecuador	0.7	0.7	0.5	0.7	0.7	0.9	0.8	8.0	0.8
Total	**32.4**	**17.7**	**25.8**	**23.0**	**52.6**	**78.1**	**80.4**	**110.7**	**192.5**

Source: Food and Agriculture Organization of the United Nations (FAO), 2017.

The production status of major quinoa producing countries is explained here under:

- **PERU:** Peru is the largest quinoa producing country in the world. As per FAO statistics (2014), quinoa production in Peru was 114300 metric tons in 2014 which is about 59.4 percent of total production in the world. The genetic materials grown in Peru are more vulnerable, but involvement of government in area expension and production policies have a favourable effect on the promotion of this crop. In year 1982–1983, the production area under quinoa in Peru was impaired due to the effects of El Niño with reported losses up to 80 percent of quinoa crop production.
- **BOLIVIA:** Bolivia is the second largest quinoa producing country in the world. Quinoa production during 2014 was 77400 metric tons which is about 40.2 percent of total production in the world. Bolivia produces more rustic varieties with a more attractive grain size that meet market demand, but with low yields which on average do not exceed 600 kg / hectare. High production levels in this country primarily reflect the intense expansion of the agricultural frontier in the last 30 years.
- **ECUADOR:** Ecuador is the third largest quinoa producing country in the world. Quinoa production during 2014 was 800 metric tons.
- **CHILE:** In Chile, Quinoa is produced on limited acreage which is reported about 1500 ha, of which 90 percent area falls in the district of Colchane in the Region of Tarapaca.
- **UNITED STATES OF AMERICA (USA):** Behind Peru, Bolivia and Ecuador, the other largest quinoa producers worldwide is United States of America. Quinoa has been cultivated in USA primarily in the high elevation San Luis Valley of Colorado where it was introduced in 1983.

  In this high altitude desert valley, maximum summer temperatures rarely exceed 30°C and night temperatures are about 7°C. The United States produces 3,000 MT annually, representing 6 percent of world production. Due to the short growing season, North American cultivation requires short maturing varieties, typically of Bolivian origin.
- **CANADA:** Quinoa is grown largely in Saskatchewan where more than 40 big quinoa growers are located. The second largest quinoa growers in Canada are Manitoba and Ontario. Now, production of quinoa is expended in Alberta where there are about 4 big quinoa growers. Canada is more variable between

30 and 1,000 MT. USA and Canada produce about 10 percent of global quinoa production.

- **EUROPE:** Small scale EU quinoa production began in 2007 in France. Presently, quinoa is grown in several countries within Europe, including France, England, Holland, Sweden, Denmark, Netherlands, Italy, Belgium, Germany and Spain, successfully on a commercial scale. The speciality of European quinoa is that it is sweet which does not contain saponin on grains. It does not require the processing for removal of saponin. The outer layer contains minerals and dietary fibres which add nutritional value over saponin containing varieties. The European quinoa market has grown rapidly in the past years. Iran, Iraq and Labenon are the countries where evaluation of quinoa varieties were conducted.
- **AFRICA:** Keniya, Algeria, Egypt, Sudan, Uganda, Ethiopia, Zambia, Ghana, Togo, Senegal, Guinea, Chad, Niger are the countries of Africa where evaluation of quinoa varieties were conducted. Keniya is main quinoa growing country in Africa.
- **AUSTRALIA:** Austrailia produces quinoa on very limited acreage.
- **CENTRAL ASIA:** The ability of quinoa to adapt to weather variations and its efficient use of water make quinoa an excellent alternative in the face of emerging challenges to food production in Central Asia. A new project called "Cross Regional Partnership for improving Food and Nutritional Security in marginal environments of Central Asia", was launched in April, 2015, Coordinated by ICBA and financially supported by the Islamic Development Bank. The overall goal of the project is to improve food and nutrition security of low-income populations living in marginal areas of central Asia through introduction of environmentally resilient and nutritionally rich quinoa. The targeted countries are Kyrgyzstan, Tajikistan and Uzbekistan.
- **CHINA:** China is major quinoa growing country in Asia.

Quinoa Export – import:

- **Exporting countries:** One of the most important destinations of Royal quinoa is export. Bolivia is the largest exporter of quinoa in the world followed by Peru and Ecuador. More than 50 percent production of Bolivia is exported to USA. The average quinoa consumption in Bolivia is estimated at 4.7 kg /person / year. Consumers in North America and Europe have a tendency to greater interest in health care, environment and social equity. The Boulder, Colorado-based Quinoa Corporation (Founded in 1983) was the first company to market the quinoa seed in the United States under the label "Ancient Harvest".
- **Importing countries:** The major importers of Bolivian quinoa grains are United States (45% of Bolivian exports), France (16%), Netherlands (13%), Germany, Canada, Israel, Brazil and the United Kingdom. Bolivia also exports quinoa flour mainly to Brazil, Chile and the Netherlands or flaked grains to Brazil, Canada and the United States. The organic niche market absorbs a significant percentage of Bolivian quinoa exports, which is marketed by companies and associated producers. There is no doubt that Bolivia leads exports mainly due

to the quality of grain produced as the "Royal" type, which does not occur the same characteristics in other countries. Peru exported quinoa grain to the United States (87%), Finland (8%) and Germany (4%).

3.5 STATUS OF QUINOA PRODUCTION IN INDIA

Quinoa was introduced in India by Late Smt. Sarvati Devi, A noted social worker of Agra district during 1975–76. She arranged seeds of quinoa in 1975 and sown in two different villages of Bichpuri Block of Agra i.e., Chhontana and Nagla Patiram under irrigated conditions. This was the first attempt of quinoa introduction / cultivation in India during *Rabi season* of 1975-76, but, unluckily, she could not get complete success due to lack of technical growing protocol.

The second attempt was made by Dr. Dharm Singh, the Author of this book, during 2003–04 and he arranged 12 demonstrations (2 in each state) on quinoa at farmer fields in Rajasthan, Haryana, Uttar Pradesh, Maharasthra, Gujarat and Delhi under "National Pilot Demonstration Programme". Due to such demonstrations, the area under quinoa was reached to about 20 hectares upto year 2004–05. The quinoa area could not increased thereafter, but, almost declined due to lack of market and complete un-awareness among farmers and consumers.

On the basis of excellent results obtained in National Pilot Demonstration Programme, Indian Medicinal Plants Marketing Federation launched a National programme entitled "Evaluation of production potential of Quinoa in marginal areas of Indian sub-continent" on 2nd October, 2009. This programme was implemented for 5 years upto 2013–14. Before implementation of this programme, the area under quinoa was almost negligible in India, but, extended remakebly to 1165 hectare with total production of 18990 quintals by the end of this programme. At present, the area under quinoa cultivation in India is 8630 hectares and total production 206257 quintals (2017–18). The average productivity of quinoa in India is presently 23.2 Q ha^{-1} (Table-3).

Table 3: *Yearwise Quinoa production in India (Year 2017–18)*

S.No.	Year	Area (ha)	Production (Q)	Average production (Q ha^{-1})
1	2009 - 10	46	483	10.5
2	2010 - 11	178	2047	11.5
3	2011 - 12	342	4412	12.9
4	2012 - 13	605	8954	14.8
5	2013 - 14	1165	18990	16.3
6	2014 - 15	3396	69958	20.6
7	2015 - 16	6278	134977	21.5
8	2016 - 17	7853	175122	22.3
9	2017 - 18	8630	206257	23.9

Source: Annual Report (2017-18) of Indian Medicinal Plants Marketing Federation,New Delhi

Table 4: *Statewise Quinoa production in India (2017-18)*

S.No.	States	Area (ha)	Production (Q)	Average Production (Q ha^{-1})
1	Andhra Pradesh	1250	32250	25.8
2	Haryana	160	4416	27.6
3	Himachal Pradesh	480	9696	20.2
4	Madhya Pradesh	375	8775	23.4
5	Maharashtra	660	17952	27.2
6	Rajasthan	1585	35821	22.6
7	Tamil Nadu	340	9146	26.9
8	Uttar Pradesh	910	21749	23.9
9	Uttarakhand	230	4899	21.3
10	Other states	2640	61553	23.3
	Total	8630	206257	23.9

Source: Annual Report (2017-18) of Indian Medicinal Plants Marketing Federation, New Delhi

At present, Rajasthan, Andhra Pradesh, Maharashtra, Gujarat, Tamil Nadu, Uttar Pradesh, Uttarakhand and Madhya Pradesh are main quinoa producing states in India (Fig. 3).

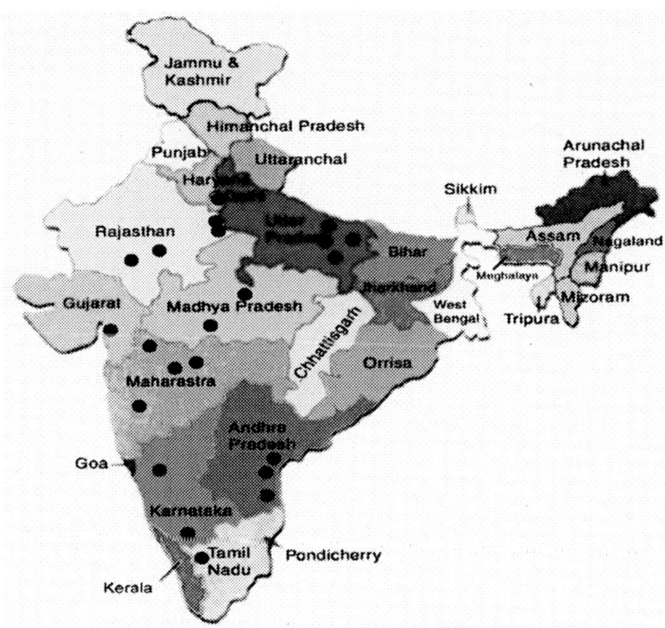

Fig. 3: *Quinoa producing states in India*

In 2013, Uttarakhand signed a Horticulture Research Agreement with Peru to promote quinoa in the state. In Rajasthan, quinoa is cultivated mainly in Jalor, Chittorgargh, Bhilwara, Jodhpur and Dungarpur districts. In Uttar Pradesh, quinoa is cultivated mainly in districts of Fatehpur, Allahabad, Kanpur, Varanasi and

Bundelkhand region. In Haryana, the area of quinoa is confined to Faridabad, Rewari, Hisar and Bhiwani districts. In Madhya Pradesh, quinoa is grown in Mandsor and Ratlam Districts. In Maharashtra, Ahmednagar, Beed, Solapur, Aurangabad, Jalna, Sangli and Pune are main districts where quinoa is being grown successfully.

In Andhra Pradesh, Anantpur district grow quinoa on large scale. In the Himalayas and the plains of Northern India, the quinoa has been established successfully with high yields.

Quinoa cultivation is spreading worldover due to rise in its demand and its establish extra-ordinary adaptability to extreme ecological conditions. Now, quinoa is being grown in more than 95 countries as per estimate of year 2017.

Chapter 4

BOTANICAL DESCRIPTION

4.1 INTRODUCTION

The scientific name of quinoa is *Chenopodium quinoa Willd* which includes the botanist abbreviation corresponding to **Carl Ludwig Willdenow.** Botanically, quinoa belongs to the **Genus** *Chenopodium,* **Class** - Dicotyledoneae and **family** – *Amaranthaceae.* Wild quinoa plants, which have a worldwide distribution, are those which have developed without human intervention and have valuable genes. Wild quinoa constitute a genetic potential that can be exploited for varietal development in the future in various parts of the planet. Some taxa and populations are characterized by tolerating and resisting insect-pest and diseases, frost and drought and also possess favourable traits in terms of nutritional value and the duration of the productive cycle.

4.2 TAXONOMIC CLASSIFICATION

Quinoa (*Chenopodium quinoa* Willd) was first botanically described by Willdenow in 1778 as a species native to South America, whose centre of origin is in the Andean of Bolivia and Peru. The classification of quinoa was first made on the basis of colour of the plant and fruits. Subsequently, it was based on the morphological types of the plant. Despite the wide variation observed, quinoa is considered to be one single species.

The botanical classification of cultivated *Chenopodium* species has been described in detail by Cronquist 1995 and Wilson 1980 and they classified quinoa taxonomically as under:

- Kingdom - Plantae (Plants)
- Sub-kingdom - Tracheobionta (Vascular plants)
- Super division - Spermatophyta (Seed plants)
- Division - Magnoliophyta (Flowering plants)
- Class - Magnoliopsida (Dicotyledoneae)
- Sub-class - Caryophyllidae
- Order - Caryophyllales
- Family - Amaranthaceae (Formerly Chenopodiaceae)
- Genus - Chenopodium
- Species - Quinoa
- Section - Chenopodia
- Sub-section - Cellulata
- Synonym - *Chenopodium album (Bathua)*

The quinoa is a member of genus *Chenopodium,* a group of about 150 species of annual or perennial herbaceous flowring plants around the world known as "goosefoots". Chenopodium is presently in the flowering plant family Amaranthaceae, although it was formerly in the family *Chenopodiaceae.* Phylogenic revision has merged the Amaranthaceae and Chenopodiaceae under the name Amaranthaceae.

The domesticated *Chenopodium species* are classified in two sub-sections i.e., Cellulata and Leiosperma. Leiosperma includes the South American species cañihua (*C. pallidculale*) and the Eurasian species group *Chenopodium album*. Quinoa (*Chenopodium quinoa*) and huazontle (*C. berlandieri sub-sp. nuttalliae*) are members of the sub-section Cellulata. Most significantly for scientists interested in quinoa improvement, there is a dearth of published information on the breeding value of related *Chenopodium species*. Another serious problem concern the taxonomy of this complex genus; for example, *Chenopodium album* has been used as a "convenient taxonomic receptacle", although this species might actually form a complex of diploids, tetraploids and hexaploids. It should be noted that a few of quinoa's close relatives have been domesticated, such that they could potentially be developed into productive cultivars in their own right, either as vegetable or as seed crops. These include *Chenopodium album* (fat hen or lambsquarters) in Eurasia, hexaploid *Chenopodium giganteum* (Khan or bithua) in the high-lands of South and East Asia, tetraploid *C. berlandieri sub-sp. nuttalliae* (huauzontle, chia or quelite) in Mesoamerica and diploid *Chenopodium pallidicaule* (cañahua) in the Andes.

The species *Chenopodium quinoa Willd* includes both domesticated and free-living weedy forms which are used either as whole plants or parts of the plant. The genus *Chenopodium* is the largest in the family Chenopodiaceae or has a World wide distribution. There is great diversity in plants and inflorescences. The genus *Chenopodium* includes about 250 species. Within the genus Chenopodium, there are several species which are cultivated as food plants, i.e., South America produce quinoa as grain crop, *Chenopodium nuttalliae* Safford and *Chenopodium ambrosioides L.* as vegetables or medicines and Mexico produce *Chenopodium quinoa Willd* and *Chenopodium pallidicaule Aellen*as as vegetable crop. The genus *Chenopodium* has been grown in various geographical areas of the world i.e., *Chenopodium album L.* in Europe; *Chenopodium giganteum D. Don or spinach* in Central Asia; *Chenopodium berlandieri Moq. Var. Nuttaliae* in Central America and *Chenopodium quinoa* and *Chenopodium pallidicaule* in South America. Also, *Chenopodium berlandieri* is distributed in North America and *Chenopodium hircinum* in the Andean and the Argentine pampa of South America.

4.3 QUINOA AND INDIAN BATHUA

Quinoa *(Chenopodium quinoa)* is morphologically very similar to Indian Bathua (*Chenopodium album)*, which is a commonly occurring weed in India and other countries. However, from agriculture point of view, these species differ with several significant traits. *Chenopodium quinoa*, the crop plant coming from South America, while *Chenopodium album* is commonly occurring in Europe and some Asian countries.

Indian bathua is kown as *Chenopodium album,* white goosefoot, pigweed, lambsquater and Bathua (Hindi). It is actually a weed that grows in waste places, with cultivated crops in tropical reasons of world and has been cultivated in Asia. *Chenopodium album* plays a pivotal role in protecting our liver, possess anti-bacterial properties, kills intestinal parasites, prevent breast cancer and treats constipation. It is considered as heart tonic. Nutritionally, its seeds are superior to wheat, rice, corn, millet and buckwheat, but, inferior to quinoa grains.

Protein is about 4.3 percent, energy 44 kcal, carbohydrade 7g, fat 0.8g, fibre 2.1g, iron 4g, calcium 280 mg, phosphorus 81 mg, vit-C 90 mg, thiamine 0.15 mg, riboflavin 0.4 mg, niacin 1.3 mg and vitamin-A 300 IU. Its protein in seeds differ from *Chenopodium quinoa* in the biological value. It was used in Poland in hard times of crop failure and hunger, as well as during war. Seeds of this species were also used for baking the so-called 'hunger-bread' and were also eaten by inhabitants of Russia, Denmark, Greece and Northern Italy.

4.4 GENETIC DESCRIPTION

Quinoa is apparently an allotetraploid, although most genetic markers, both morphological and molecular, segregate in a typical disomic Mendelian fashion. The chromosome number of cultivated quinoa (*Chenopodium quinoa Willd*) is 2n = 4X = 36. The basic chromosome number for all known species in the genus and most species in this family is X=9 and quinoa has numerous wild relatives with chromosome numbers of 2n=18, 36 and 54, indicative of its apparent tetraploid origin.

4.5 NATURE AND GROWTH HABIT

Quinoa is an annual plant; usually herbaceous dicot that reaches to a height of 0.5 to 1.5 m, depending on environmental conditions and genotype. Many cultivars reaching upto 3 m in height, generally with large number of branches and a big leaf size. Quinoa is a plant species that has not been improved by plant breeding, thus, shows significant variability. The plant shows good growth in India with many cultivars reaching upto 1.5 m in height, generally with large number of branches and a big leaf size. Quinoa plants can display a variety of colours ranging from green to red and purple with many intermediate colours in between. The branching habit is more common in races cultivated in the Andean valleys of Southern Peru and Bolivia, where as the simple habit is observed in a few races grown in the Altiplano and many of the races of central and northern Peru and Ecuador.

4.6 MORPHOLOGY OF QUINOA PLANT

- **Stem:** The main stem of quinoa is robust, with scaly and reddish bark. It may be branched or without branching depending on the ecotype, race, planting density and the environmental conditions in which it is grown. Inside, the hollow stem is spongy and rich in pectin and cellulose that can be used for the manufacturing of paper and paper board. In the early stages of growth, stem is massive and soft, but it becomes hollow and spongy as maturity approaches. In contrast, the cortex is firm and compact. The outer stem can be green, green with red axil, green with purple or red strips or red. The cross section of stem is circular in the area near the root, becoming angular at the height of the branches and leaves.

The branches originate from the exils of each leaf on the stem. Their length may vary from just a few centimeters to the same length as the main stem, according to cultivar and environmental conditions.

- **Roots:** Quinoa has a well-developed and fibrous roots with, highly ramified tap-root system. The tap root divides just below the root-neck giving rise to secondary and tertiary roots. When the root is fully developed, it can reach or penetrate in the soil as deep as 1.8 m below the surface which makes quinoa plants more resistant to drought. The root system develops from a tap root to form a highly branched system that makes plants more resistant to drought.

Fig. 4: *Quinoa plant*

- **Leaves:** Quinoa bears alternate leaves of various colours ranges from green to red, through yellow and violet, depending on the nature and presence of the pigments. The leaf colour is due to the presence of beta-cyanins. At maturity of the plant, leaves turn yellow or red, depending on variety.

  The leaves exhibit polymorphism; the upper leaves (usually the top ones) being rhomboid or triangular, while the lower leaves (around the inflorescence) can be lanceolate. The edge of Leaves may be dentate (toothed) with up to 43 teeths on their margins or smooth and they contain granules on their surface giving the appearance of being covered with sand. These granules contain cells rich in calcium oxalates which allow plants to absorb and retain moisture better from environment, while protecting it from frost. Calcium oxalates are able to retain a water film, which increases the relative humidity of the atmosphere surrounding the leaf thereby reducing transpiration.

- **Flowers:** Quinoa plants are gyno-monoecious which bears hermaphrodite (both male and female organs) and unisexual female flowers. Some cultivars show male sterility in some or all female flowers. Quinoa plants bear both pistillate and perfect forms.

Quinoa flowers are incomplete, sessile and have no petals. At bloom, they are tiny, reaching a maximum size of 3 mm. The flowers are very small and dense, positioned in groups forming glomerules, which are sessile and make emasculation difficult. The hermaphrodite flowers are located at the distal end of glomeruli and bear five perianth lobes, five anothers and a superior ovary with two or three stigmatic branches. Style is characterized by 2 or 3 feathery stigmas. The pistillate flowers are at the proximal end. Some cultivars show male sterility in some or all female flowers. The flowers remain open from 5 to 7 days. As they do not open simultaneously, the duration of flowering has been determined as 12 to 15 days.

- **Inflorescence:** The inflorescence of quinoa is racemose, called a panicle, because it has a more developed main axis which gives rise to secondary and in some cases tertiary axes. The Panicle has 15–70 cm length, developed from the top of the plant and in the axils of lower leaves. Each panicle has a central axis from which a secondary axis emerges. The inflorescence in quinoa is of two types i.e., amaranthiform and glomerulate or intermediate according to their panicle shape. The name amaranthiform owing to its resemblance to the inflorescence of the genus *Amaranthus*. In amaranthiform type of inflorescence, glomeruli (short branches bearing a group of flowers or grains) are inserted directly on second order axes. In glomerulate type, small groups of flowers (glomeruli) originate from tertiary (third order) axes.

  Panicle shape is genetically determined by a pair of genes, the glomerulate form being totally dominant over the amaranthiform and it therefore seems doubtful to classify intermediate panicles. The terminal panicle may be defined (completely differentiated from the rest of the plant) or branched (when there is no clear differentiation because the main stem has relatively long branches that give the panicle a distinctive conical shape). The panicle may also be loose or compact, which is determined by the length of the secondary axes and pedicels, being compact when both are short.
- **Pollination:** Quinoa is usually self-pollinated as the green hypogynous flowers have simple periath. However, about 10 to 15 percent cross pollination is recorded in quinoa. It not require pollinators for pollination.
- **Fruits:** The fruit may be conical, cylindrical or ellipsoidal with a coating of thin pericarp or a layer of saponins. This layer contain 4 percent saponin which gives a bitter taste and foamy aspect in cooking war. The function of this saponin is to protect grains from predators. It must be removed when the seed is intended for human consumption, because, saponin can cause serious health problems like destroying red blood cell membrane. Fruits are an achene, comprising several layers viz., perigonium, pericarp and episperm. Episperm is composed of four layers. The fruits can reach up to 2.66 mm in diameter according to the variety. The outer layer determines the colour of the seed, contains saponin and may be removed with water.
- **Seeds:** The seeds are generally flat, oval-shaped which are produced in large clusters on a panicle. Seed size and colour are variable. The quinoa seeds

measure about 2 mm in diameter. Seeds can be black, red, pink, orange, yellow or white in colour depending upon the cultivar type. The seed colour is due to resinous coating that contain 2 to 6% saponins. Black seeds are dominant over red and yellow, which in turn are dominant to white seed colour. Seed weight ranges from 2 to 3 mg.

4.7 SEED BIOLOGY

- **Seed structure:** The episperm surrounding the grain is composed of four layers. The outer layer which determines the colour of the seed is rough, brittle, easily removed with water and contains bitter substances known as saponins, which protects grains from birds due to toxic and bitter taste. It is necessary to eliminate saponins from the seeds before eating or processing for the manufacture of food products. Seed covered by epidermis represent 60 percent part of the seed. It is most rich in starch. Epidermis is developed by the embryo from which yelloyish white tail like structure, called radical, stands out. The radical constitutes 30 percent of total seed volume wrapping with a curature of 320 degree.

In quinoa, the main storage reserves is clearly divided in 3 main storage components which can be distinguished in the mature seeds as described below:

- A tissue that stores mainly starch, which is the perisperm in quinoa.
- An embryo that store principally protein and lipids which is the richest part of the seed in proteins (35 – 40).
- Non-embryonic tissue that stores proteins, lipids and minerals which is the aleurone layer in gramineae and the micropilar cone in quinoa.

- **Seed Embryo:** The embryo comprises 60 percent of volume within the pericarp and these results in the higher protein content of the seed in comparision to cereal grains. This embryo is the richest part of the seed in protein (35 to 40%).
- **Seed germination:** Quinoa seeds do not exhibit dormancy. Quinoa seeds have a notably short germination period. When adequate moisture is present, germination occurs within 24 hours after planting and seedlings emerge in 3 to 5 days. Quinoa seeds may not germinate if conditions are warm and may need to be refrigerated for a week to vernalize them to obtained adequate germination.
- **Seed dormancy:** Quinoa seeds do not have dormancy. Seeds may remain viable in the soil for two or three years without germination and germinate immediately when conditions became suitable.
- **Grain quality:** Quinoa seeds are very nutritious and content of single component vary. Seeds in comparision with other cereals are higher in protein content (average - 14.6%). Its protein fraction is well balanced comes close to the ideal protein, qualitatively corresponding to casein. It has more favourable ammino acid composition also – higher content of lysine, methionine and threonin and high content of arginine and histidine which are important in nutrition of infants.

The starch content is approximately 60 percent represented by small granules exploitable in food industry as an ingradient in cream substitute. Quinoa starch shows better viscocity than other cereals. Low content of amylose and the size of starch granules are significant factors affecting the uses of quinoa flour. Products are much elastic and nutritive by 10 percent addition of quinoa in common flour. If higher amounts of quinoa flour is added than the volume and porocity of paste is lower and consistency is tough. Amount of lipids in quinoa is about 8 percent of dry mass. The oil is very stable due to high anti-oxidants. Oil contains 54 percent of oleic acid. Quinoa is a good source of thiamine, folic acid and vitamin-C, but has lower content of niacin (B3) in comparison with other cereals. Seeds contain more Ca, P, Mg, Fe, Zn, Na and Cu than cereals. It is very important to realise distribution of every single mineral in fractions of the seed. The most significant losses caused by a braiding are on Na, Fe and Mn. Seeds contain some anti-nutritious compounds, such as saponins, phytates, tannins and protease inhibitors. Of all these compounds, the amount of saponins causing bitter taste is the least desirable. Therefore, quinoas are sorted out in three ecotypes – bitter, medium and sweet. Sweet type have usually white and bigger seeds.

- **Reproductive biology:** Type of reproduction in quinoa can be both self and cross pollinated. Quinoa has a sexual reproduction process. Sexual reproduction is a biological process in which meiosis and pollination result in genetic variation which is exploited in genetic crop breeding. All the flowers in quinoa's inflorescence open within 12 to 15 days from the onset of flowering.

Chapter

5

GENETIC DIVERSITY, SPECIES AND VARIETIES OF QUINOA

5.1 GENETIC DIVERSITY

The nearest wild species to *Chenopodium quinoa* are *Chenopodium hircinum* and *Chenopodium berlandieri* which have the same number of chromosomes (2n = 4x = 36) and *Chenopodium pallidicaule* with 2n = 2x =18 chromosomes. Sympatric wild population of domesticated quinoa exist under cultivation and morphological and electrophoretic similarities can be noted between one and the other in each locality, which indicate that domesticated quinoas are generally accompanied by wild populations in their various distribution areas.

The Andean Region is considered one of the eight centres of crop origin and diversity. This is where there is the greatest genetic diversity of both wild and cultivated quinoa which can still be found in the wild and in the Andean farmer's fields. Among Andean crops, quinoa has received most dedication and support in Ecuador, Peru and Bolivia. Assessments of the genetic variability available has allowed quinoa categorize into five major groups according to their adaptations and some highly heritable morphological traits, which are easily detectable and which are maintained throughout the whole area of distribution.

There are about 120 species of quinoa, but only 3 are cultivated. Cultivated / cooked quinoa grains have a fluffy, creamy, slightly crunchy texture and are very nutritious and delicious with a somewhat nutty flavour. Cultivated quinoas have a high genetic diversity, showing variability in the colour of the plant, inflorescence and seed which confers them with a wide adaptability to different ecological conditions (soil, rainfall, temperature, altitude, resistance to frost, drought, salinity or acidity).

Genetically and ecologically, quinoa varieties can be separated into two economically important sub-groups:

- **Andean ecotypes**: Adapted to the high altitude environments of the Altiplano regions of Bolivia and Peru.
- **Coastal ecotypes:** Adapted to the humid, disease prone coastal low lands of Chile and Ecuador.

Recently, several modern, improved varieties have been released, predominantly within the Andean sub-group. These varieties are genetically uniform inbred lines, and they display substantially higher yields than their land race counterparts.

The adaptation of certain quinoa varieties is possible even under marginal environments for the production of seeds with high protein and mineral content. Quinoa's aptitude to produce high-protein grains under ecologically extreme conditions makes it important for the diversification of agriculture as in high-altitude regions of the Himalayas and North Indian Plains. Cultivated quinoa exhibit great genetic diversity showing variability in the colouring of the plant, inflorescence and seeds, type of inflorescence, protein content, saponin content, Betacyanine and calcium oxalate crystals in the leaves, so that a wide adaptation to different agro-ecological conditions may be seen (soils, precipitation, temperature, altitude, resistance to frost, drought, salinity or acidity).

Quinoa may be considered an oligocentric species from the point of view of its variability. With an ample centre of origin of multiple diversification, the Andean region and within it, the shores of Lake Titicaca, are where the greatest diversity and genetic variation occur. The risk of genetic erosion are due to not only to the loss of viability in gene Banks (at present it exceed 15 per cent annually), but also that occurring in the areas of diversification, particularly in places where cultivars and modern varieties are promoted for commercial purposes. The case of wild species is even more delicate, with small, isolated populations which have not been collected in time and which are inexorably vanishing. There are also risks from environmental factors and natural disasters. Insitu conservation is an alternative, although there are difficulties of a socio-economic nature and also regarding institutional support.

Types of Quinoa

According to taste

According to tase, quinoa is differentiated in two categories i.e., Bitter quinoa and Sweet quinoa. Bitter quinoa is the most valued and so much popular in the markets because of their large size of the grains. It is relatively resistant to frost and drought, and can be grown in rough climatic conditions. This type of quinoa is known as *Royal Quinoa.* The grains of quinoa of this type contain a high level of saponin (more than 0.11%) that provides it a bitter taste. This saponin has to be removed before any type of consumption.

The another type is sweet quinoa which is free from saponins or having less than 0.11 percent. Sweet quinoa not require any type of processing like bitter quinoa.

According to the colour

According to the colour, quinoa is differentiated mainly in 4 categories i.e., White, Red, Black and multi-coloured as explained hereunder:

- **White quinoa:** It is most common kind of quinoa havig white or light golden grains. This type of quinoa is available in most of the local super markets, organic food stores and health food stores. It is some time called *ivory quinoa* having very mild, subtle flavour. It is the mildest, least crunchy, sweeter than other coloured quinoa varieties and similar in taste to white rice. White grains are extremely easy and fast to cook by mixing 1 part quinoa to 2 parts of water for 10–15 minutes. The grains of white quinoa are used in different kind of dishes, such as salads, soups, burgers etc. The white grained varieties have greater plant height, length of panicle and number of glumerulates.

Fig. 5: *White quinoa grains*

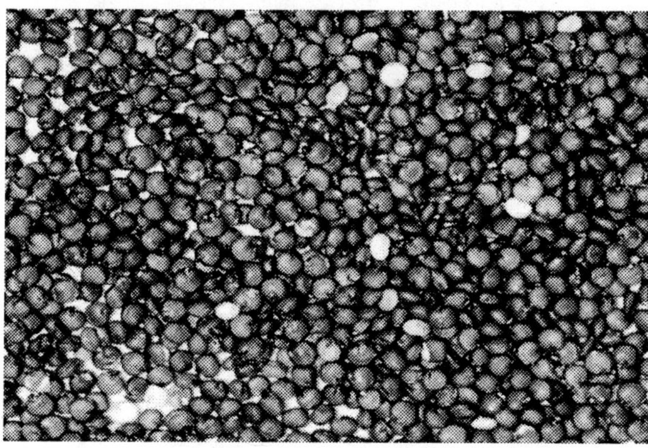

Fig. 6: *Red quinoa grains*

- **Red quinoa:** Red quinoa is most popular among Inca warriors and they worship red seeds of quinoa. Red colour of grains ranges from vivid orange to dark red, turns brown and loses its bright colour when cooked. Red quinoa has a marvelous earthy and fruity favour. The grains of red quinoa have higher levels of phenolic compounds. It retains its shape (better than white quinoa) after cooking and hence it is better for use in cold salads or other recipes where a distinct grain is especially desirable.
- **Black quinoa:** Black quinoa is a hybrid variety that was developed by farmer in the Colorado Rockies. This happened when quinoa seeds were cross-bred with the lamb's quarter, a type of wild spinach. Black colour is due to anthocyanins which are a class of compounds that prevent UV damage and oxidation of the plants. It is varied in taste from other types of quinoa, having a crunchy texture even remains after it is cooked although the colour turns

inky. Black quinoa is slightly fibrous and it cooks fast. Quinoa of this type keeps its striking black colour when cooked and is sweeter than the white quinoa. Like red and white quinoa, it is a fluffy texture when cooked and has a slightly nutty flavor which makes it an alternative to rice or couscous. It is crunchiest of the three i.e., white, red and black. Black quinoa keeps its striking black colour when cooked.

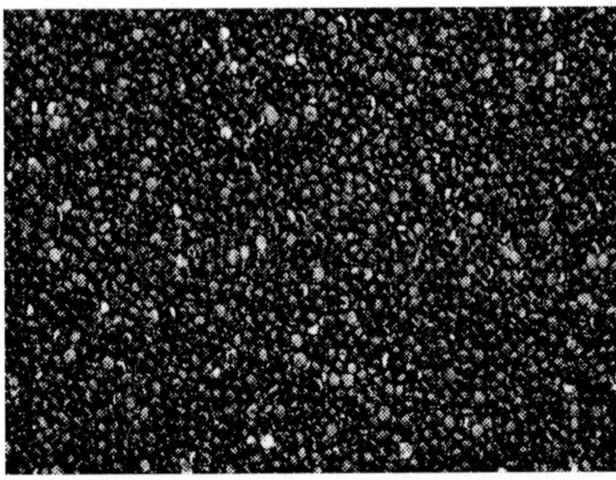

Fig. 7: *Black quinoa*

Fig. 8: *Multi-coloured quinoa*

- **Multi-coloured quinoa:** Multi-coloured quinoa is similar to white, red and black in respect to nutritional value, but makes for a more attractive appearance once cooked.

According to ecotypes:

Quinoa grows in the altitudes from the sea level to the Andean high lands. On the basis of this nature, there are five groups of quinoa.

- **Sea level Quinoas:** Found in the areas of Linares and Concepción (Chile) at 36° South latitude and 500° altitudes. They are more or less robust plants, 1.0 to 1.4 m tall, with branched growth and produce transparent cream-coloured grains. These quinoas have great similarities with

Chenopodium nuttalliae that is grown only in Mexico at 20° North Latitude. The plants of this category are tall, branched and long day type with yellow and bitter seeds.

- **Valley Quinoas**: Quinoa plants of this category adapt between 2500 to 3500 metres in Andease valley. They are characterized by their height up to 2.5 m or more and are greatly branching with lax inflorescences and are usually resistant to mildew disease caused by *Peronospora farinosa.*
- **Altiplano Quinoas:** They develop in larger areas as monocultures and between 3600 and 3800 masl, in the area of the Peruvian-Bolivian Altiplano. This is the area in which the greatest variability in characters is found and the grains with the most specialized uses are produced. The plants grow to heights of 0.5 to 1.5 m, with a stem ending in a generally compact main panicle. This group contains the largest number of improved varieties and also those most susceptible to mildew when taken to wetter areas.
- **Salt-flat Quinoas:** They grow in salt-flat areas to the south of the Bolivian Altiplano. The driest area has 300 mm of precipitation. They are grown as a single crop with spacing of 1 m x 1m in holes to make better use of the little humidity available. These quinoa have the largest grain size (>2.2 mm in diameter) and are known as "Royal Quinoa". The grains are characterized by a thick pericarp and high saponin content.
- **Quinoas of the Yungas:** A small group of quinoas that have adapted to the conditions of the Bolivian Yungas at altitudes between 1,500 and 2,000 masl and are characterized by a somewhat branched development. They are green plants which reach heights of up to 1.20 m and when they are in bloom, the whole plant takes on an orange colour.

Varieties of Quinoa

- **Varieties of Bolivia:** 22 varieties are obtained in Bolivia by genetic improvement through hybridization and selection. There is also a significant complex of bitter varieties known as "Royal Quinoa" which includes several local races: Royal white, Mañiqueña, Huallata, Toledo, Pink Mok'o, Tres Hermanos, K'ellu, Orange Canchis, Pisankalla, Pink Pandela, Perlasa, Achachino, Hilo, White rose, Mok'o, Timsa, Lipeña, Chillpi Amapola, Pink Chillpi, Utusaya and Pink Canchis. Sajama, Real Blanca, Chucapaca, Kamiri, Huaranga, Pasancalla, Pandela, Tupiza, Jachapucu, Wila coymini, Ǩellu, Uthusaya, Cullpi, Kaslali and Chillpi.
- **Varieties of Peru:** Yellow Maranganí, Kancolla, Juli White, Cheweca, Witulla, Salcedo-INIA, Iplla-INIA, Quillahuaman INIA, Camacani I, Camacani II, Huariponcho, Chullpi, Coporaque Red, Ayacuchana-INIA, Huancayo, Hualhuas, Mantaro, Huacataz, Huacariz, Yanamango Pink, Namora, Tahuaco, Wilacayuni, Pacus, Junín Pink, Junín White, Acostambo, Ayacuchana White, Yocara, Blanca de juli, Amarilla, de Marangam, Rosada Blanca de Junin, Hualhuas, Blanca and Chucapaca.
- **Varieties of Ecuador:** Tunkahuan, Ingapirca, Cochasqui, Imbaya, Chaucha, Tanlahua, Piartal, Porotoc, Chimborazo bitter, Imbabura bitter and Purple.

Tunkahuan variety is effective and is the most planted in the high lands of Ecuador, Peralta also indicates that in 2004 the variety Pata de Venado was released. In Colombia, the varieties Nariño and Quitopampa sweet were obtained, while Chile produced the varieties Baer, Lito, Faro and Picchaman andArgentina-the Jujuy White.

- **Varieties of Colombia:** Dulce de Quiopampa.
- **Varieties of Argentina:** Blanca, de Jujuy.
- **Varieties of Chile:** Baer, Lito, Faro and Pichaman.

5.4 GERM PLASM BANK IN THE ANDEAN REGION

In order to protect the vast phenotypic and genotypic variability of quinoa of the Andean Region, since the 60's germplasm banks have been implemented throughout the Region, with universities and agricultural entities in charge of their management and conservation. The number of accessions stored in the Region is variable, Bolivian and Peruvian entities holding the greatest quinoa variability in their gene banks. According to available information, five countries in the Andean Region have germplasm banks where quinoa is preserved along with other Andean crops.

In Bolivia, there are six entities with gene banks which conserve more than 5,000 quinoa accessions:

- The Toralapa Centre of the National Institute of Agricultural and Forestry Innovation – INIAF.
- The Choquenaira Experimental Station of the Universidad Mayor de San Andrés – UMSA.
- The Biotechnology and Phytogenetic resources research centre (CIBREF) of the Technical University of Oruro (UTO)
- The Tiahuanacu Academic Unit of the Bolivian Catholic University – UCB.
- The Kallutaca Experimental Centre of Public University El Alto – UPEA.
- The Centre of Comunal Research and Promotion– CIPROCOM.

Bolivia´s main quinoa germplasm collection is preserved in the INIAF, Toralapa Centre and forms part of the National Germplasm Bank of Andean Grains. It is home to a wide genetic variability and currently maintains 3121 accessions, both cultivated and wild, which were gathered from communities in the Altiplano and the valleys in the Bolivian departments of La Paz, Oruro, Potosi, Cochabamba, Chuquisaca and Tarija. It also has germplasm from Peru, Ecuador, Colombia, Argentina, Chile, Mexico and the United States.

Chapter 6

PRIMARY METABOLITES AND NUTRITIONAL VALUES

6.1 INTRODUCTION

Metabolites are compounds synthesized by plants for main two type of functions i.e., Essential functions, such as growth and development and specific functions, such as defence against herbivory, pollinators attraction etc. The metabolites which perform essential functions in the plants are known as primary metabolites. The Primary metabolites are also called "Central metabolites". They directly involve in growth, development and reproduction, and they are essential for survival or existence of all kind of floweing plants. Primary metabolites comprise many different types of organic compounds i.e., protein, carbohydrates, lipids, nucleic acids, vitamins and minerals. The absence or deficiency of primary metabolites in the plants can cause the death of plants depending upon severity.

6.2 PRIMARY METABOLITES

The primary metabolites of quinoa plants have been explained here under:

- **Proteins:** Grains, leaves and inflorescences of quinoa plants are good sources of high quality protein. Quinoa grains contain 13.8 – 21.9% protein (average 16.3% on dry basis) depending on the variety. Its protein is constituted by water soluble albumins and globulins with 80% of digestibility. Quinoa protein is gluten free, complete and rich in lysine (5.1–6.4%) and methionine (0.4–1%) contents. Compared to cereal grains, the total protein content of quinoa (16.3% dry basis) is higher than that of barley (11%), rice (7.5%) and corn (13.4%) and is comparable to that of wheat (15.4%). It has also been found that quinoa leaves have a high content of quality proteins.

It is evident from research findings that despite of good nutrient content in quinoa, the amino acids in the raw unwashed flour are not fully available because they contain substances that interfere with the biological utilization of nutrients. These substances are called saponin glycosides. After roasting, the quinoa grains acquire a brown colour caused by the presence of reducing sugars that produce a Maillard reaction. The lysine in this form is not biologically useful (it loses its nutritional value).

Protein of any source is consisted with amino-acids, which are small units that make up proteins and have vital functions in the body. Amino acids are of following three types:

A. ESSENTIAL AMINO ACIDS

The term essential denote that our body cannot produce them and need to get them from the diet. There are 9 essential amino acids which cannot be synthesized by human body. They are lysine, phenylalanine, threonine, tryptophan, isoleucine, methionine, valine, histidine and leucine. If protein of any food contains all these essential amino acids, it is called a complete protein. Most of the foods are deficient in certain essential amino acids. The biological functions of essential amino acids have been summarised below:

- ***Lysine:*** This essential amino acid is deficient in most cereal grains like wheat, maize, rice etc while quinoa proteins are particularly high in lysine. It is building blocks of protein and plays an essential role in the production of Carnitine, a nutrient responsible for converting fatty acids into energy and helps in lowering cholesterol. Lysine amino acid enhances immune function by aiding in the formation of antibodies, promotes gastric function and assists in cell repair. Lysine is involved in the metabolism of fatty acids, helps calcium absorption and transport and even seems to slow or prevent cancer metastasis together with vitamin-C.
- ***Phenylalanine:*** It is naturally found in breast milk of mammals. Quinoa is rich in phenylalanine, a brain stimulant and the main element of the neuro-transmitters that promotes alertness and relieves pain and depression. It is used in manufacture of food and drink products as nutritional supplements for its analgesic and anti-depressant effects. It is a direct pre-cursor to the neuromodulator phenethylamine, a commonly used dietary supplement.
- ***Threonine:*** This essential amino acid involved in liver detoxification, participates in the formation of collagen and elastin and facilitates the absorption of other nutrients. It treats various nervous system disorders including spinal spasticity, multiple sclerosis, familial spastic paraparesis and amyotrophic lateral sclerosis.
- ***Tryptophan:*** This essential amino acid act as building blocks in bio-synthesis of proteins. Proteins are required to sustain life. Tryptophan is immediate precursor of the neuro-transmitter serotonin, for which it has been used successfully in cases of depression, stress, anxiety, insomnia and compulsive behaviour. It is needed to prevent illness or death. Tryptophan content of quinoa is similar to that of wheat, higher than that of other cereals i.e., barley, pearl millet and rice.
- ***Isoleucine:*** This essential amino acid is not synthesized in human body and must be ingested through our diet. It is synthesized from pyruvate employing leucine bio-synthesis enzymes in other organisms, such as bacteria. It is associated with the disease called Maple Syrup Urine Disease which results in discolouration and a sweet smell in the patient's urine. Isoleucine of quinoa participate in the production of muscle energy, improve neuro-muscular disorders, prevent liver damage and balance blood sugar levels.
- ***Methionine:*** Human body cannot synthesize this essential amino acid. It is used by the liver to produce s-adenosine-methionine, a substance particularly

effective in treating liver disease, depression, osteo-arthritis, brain disorders, fibromyalgia and chronic fatigue, among other ailments. It also acts as a powerful detoxifying agent that significantly decreases the levels of heavy metals in the body and exerts a strong protection against free radicals.

- ***Valine:*** Quinoa is good source of valine amino acid containing about 42.1 mg in each gm. It is a branched-chain amino acid providing a stimulant activity. It is synthesized in plants via several steps starting from pyruvic acid. Valine as well as other branched-chain amino acid are associated with insulin resistance, as higher levels of valine are observed in the blood of diabetic patients. In humans, a protein restricted diet lowers blood levels of valine and decreases fasting blood glucose levels. This amino acid prevent the breakdown of muscles with an extra glucose responsible for energy production during physical activity. Valine is also a precursor in the penicillin bio-synthetic pathway and is known for inhibiting the transport of Tryptophane across the blood brain barrier. It is essential for smooth nervous system and congnitive functioning. It is important for every day body functions and for maintaining muscles as well as for regulation of immune system.
- ***Histidine:*** It is most common amino acid in protein highly required by the body especially in the growing stage of children. It helps in transport and combine metals as well as iron and copper. It also enhances absorption of calcium, decreases the levels of histamine resulting in the control of diarrhea. Excessive histidine cause constipation. Low or deficiency of histidine is correlated with rheumatic arthritis and stomach / duodenal ulcer in adults and cause eczema in infants. It triggers the discharge of the digestive substances that stimulates the production of gastric juices called gastrin. Without sufficient histidine in diet, good digestion may possibly be impaired.
- ***Leucine:*** It is a dietary amino acid which directly stimulate the synthesis of muscles protein. It had been found to slow the degradation of muscle tissue by increasing the synthesis of muscle proteins. It increases muscle mass and helps muscles recovery after exercise. It also regulate blood sugar and supplies the body with energy. These functions make it invaluable when the body is stressed.

B. CONDITIONALLY ESSENTIAL AMINO ACIDS

Other six amino acids are considered conditionally essential in the human diet, meaning their synthesis can be limited under special patho-physiological conditions, such as pre-maturity in the infant or individuals in severe catabolic distress. These six are arginine, glycine, cysteine, glutamine, tyrosine and proline.

- ***Arginine:*** It is used in bio-synthesis of proteins and considered as a semi-essential or conditionally essential amino acid. Arginine plays an important role in cell division, the healing of wounds, removing ammonia from the body, immune function, promote production of growth hormones, repair and protect muscles, and detoxify the liver. Some evidence shows that arginine may help improve blood flow in the arteries of the heart.

- ***Glycine:*** It is conditionally essential amino acid. Its principal function is as a precursor to proteins. It acts as inhibitory neuro-transmitter in central nervous system especially in the spinal cord, brain stem and retina and act as a regulator of motor function. It can improve sleep quality.
- ***Cysteine:*** It is conditionally essential amino acid which protects the liver by binding to heavy metals and facilitating their removal in addition to destroying free radicals and boosting the immune system.
- ***Tyrosine:*** It is considered as non-essential (conditionally essential) amino acid and body makes it from another amino acid called pheylalanie. It is an essential component for the production of several important brain chemicals called norepinephrine and dopamine. It has an important anti-stress effect and plays a fundamental role in alleviating depression and anxiety, among other functions.
- ***Glutamine:*** It can be synthesize in the muscles of human body making it as non-essential amino acid. It is distributed by the blood to the organs that need it. It might help gut function, the immune system and other essential processes in the body especially in times of stress. During synthesis of glutamine to glutamic acids, the brain is protected from the toxic effects of ammonia. This protection against the cell toxins prevents impediments to brain functions and improves long and short term memory. It is involved in the production of energy for the brain and important phenomena, such as learning, memory and neuronal plasticity.
- ***Proline:*** It is formally not an amino acid, but an imino acid. Nonetheless, it is called an amino acid. Proline is an alpha amino acid that is used in bio-synthesis of protein. It does not have a hydrogen when it is in a peptide bond. Proline is unique in that it is only the amino acid where the side chain is connected to the protein to the protein backbone twice forming a five-membered nitrogen containg ring.

C. NON-ESSENTIAL AMINO-ACIDS

An amino acid that can be produced by human body and is not required in the diet, are called non-essential amino acids. There are 5 non-essential amino acids i.e., Alanine, Asparagine, Serine, Aspartic acid and Glutamic acid. With regard to "non-essential" amino acids, quinoa contains more than triple amount of triple histidine as compared to wheat. This substance is essential for infants because the body cannot synthesize it until adulthood, so it is strongly recommended that children acquire it through their diet, especially during periods of growth. It also has a slightly anti-inflammatory action and participates in the immune system response.

- ***Alanine:*** It is non-essential amino acid and energy source for the muscles, brain and nervous system. Glucose-alanine cycle is the main function and helps to balance glucose and nitrogen in the body.
- ***Serine:*** It can be synthesize in human body under normal physiological circumstances, making it a non-essential amino acid. It is potent natural moisturizer. It participate in the biosynthesis of purines and pyrimidines.

It is precursor to several amino acids including glycine and cysteine and tryptophane in bacteria.

- ***Asparagine:*** Asparagine is known for its key role in the biosynthesis of glyco–proteins. It is also essential for synthesis of other proteins. Human nervous system also needs this amino acid to be able to maintain an equilibrium. It is required for development and function of brain. It also plays an important role in the synthesis of ammonia.
- ***Aspartic acid:*** This amino acid is present in quinoa, but uncommon in other cereals. It improves liver function and is essential for the maintenance of the cardio-vascular system.
- ***Glutamic acid:*** It is essential for protein synthesis. It offers several health benefits and maitains the healthy functioning of the body. It supports good health of immune system, digestive system and aids in the energy production in the body. It is most common neuro-transmitter which is found in the brain and spinal cord. It plays integral role of excitatory neuro-transmitter in the central nervous system as it helps in the regulation of number of ions like sodium, calcium and magnesium.

The literature on human nutrition indicates that only four amino acids are likely to limit the quality of mixed human diets. These amino acids are lysine, methionine, threonine and tryptophan. Quinoa grains contain all essential amino acids in excellent proportions. Due to this reason, quinoa is considered an excellent source of protein including lysine.

The most important amino acids present in quinoa protein and other traditional foods are presented in Table 5.

Table 5: *Composition of Amino acids of quinoa and cereals (Barley, soybeans and Wheat)*

Amino acids	Quinoa	Barley	Soy-bean	Wheat
	Mg /g protein			
Essential				
Histidine	28.8	–	26.7	23.5
Isoleucine	35.7	–	44.5	43.2
Leucine	59.5	–	72.0	82.8
Lysine	54.2	37.2	57.8	36.2
Methionine	21.8	19.2	10.6	23.5
Phenylalanine	42.0	56.1	49.2	53.5
Threonine	29.8	34.0	38.6	35.8
Tryptophan	11.8	16.6	12.2	11.5
Valine	42.1	49.0	47.6	61.1

Non-essential				
Alanine	41.6	39.0	42.2	58.0
Serine	40.2	42.2	50.0	52.6
Aspartic acid	80.3	62.5	136.3	94.0
Glutamic acid	132.1	261.2	151	195.1
Conditionally essential				
Arginine	77.3	50.1	69.5	–
Cysteine	14.4	22.1	12.1	20.5
Glycine	49.2	36.2	38.6	45.5
Tyrosine	18.9	28.7	36.2	33.4

D. SPECIALITY OF QUINOA PROTEIN

The main speciality of quinoa protein is free from gluten. Gluten is a substance present in certain grass family grains like wheat, which may induce stomach upset and diarrhea with gluten sensitivity or celiac disease. Quinoa's exceptional amino acid richness confers it with very interesting therapeutic properties. Quinoa protein have a balanced composition of essential amino acids just similar to the amino acid composition of milk protein casein.

DIGESTIBILITY OF QUINOA PROTEIN

The digestibility or bio-availability (true digestibility) of protein of quinoa varies depending on the variety.

The studies on protein digestibility indicate that the values of true digestibility of protein is placed in three ranges:

- **High digestibility**: From 93 to 100%. Mostly animal foods and soy protein isolate comes under this category.
- **Intermediate digestibility**: From 86 to 92%. The protein of polished rice, whole wheat, oat meal and soy flour come under this category.
- **Low digestibility**: From 70 to 85%. The protein of corn and different types of pulses including beans and lentils come under this category.

According to this classification, protein of quinoa grains come in third category i.e., with low digestibility.

Protein Efficiency Ratio (PER)

Protein Efficiency Ratio is the ratio of gm is based on the weight gain of a test subject divided by its intake of a particular food protein during test period.

$$\text{PER} = \frac{\text{Gain in the body mass (g)}}{\text{Protein intake}}$$

The protein efficiency ratio of cooked quinoa is 30% greater than that of uncooked quinoa.

E. CARBOHYDRATES

The main carbohydrates of quinoa are starch (58 to 68%) and sugars (5%), which make it an ideal source of energy. Quinoa releases energy slowly into the human body owing to its high dietary fibre content. It plays a basic nutritional function and may have different physiological health effects i.e., provision of energy, effects on gastric emptying, control of blood glucose and insulin metabolism; protein glycosylation; cholesterol and triglyceride metabolism. The carbohydrate content of quinoa is comparable to that of barley and rice (Table - 7).

Carbohydrates of quinoa can be classified into following three principal groups, according to their degree of polymerization:

- **Starch:** Starch is the major component of carbohydrate in all food grains and major energy reserve in plants. The most important sources of starch are cereal grains, legumes and tubers. Quinoa grains contain 58 to 68% starch. The content of starch in red, yellow and white cultivars of quinoa is reported to be 59, 58 and 64%, respectively. The starch of quinoa has a higher swelling power than that of barley, wheat, rice, amaranth and potato. Starch is consists of two polysaccharides i.e., amylose (14.3–27.7%) and amylopectin - 77.5%.

  In the plants, the starch is present in granular form. The granules of quinoa starch have a diameter of 2 μm, being smaller than starch of the common grains. Genetic variation of starch granule size in the Bolivian quinoa collection ranged from 1 to 28 μm. Starch granules occur in the perisperm cells. The digestibility of starch of raw quinoa is reported at 22%, while that of autoclaved, cooked and drum-dried seeds was 32%, 45% and 73%, respectively. The saponins present in quinoa seeds did not affect the digestibility of the starch.

- **Sugars:** In addition to polysaccharides (monosaccharaides, disaccharides and polyols), the quinoa grains contain free sugars in small quantity i.e., about 5%. The individual sugars present in quinoa are mostly maltose, followed by D-galactose and D-ribose and it also contains fructose and glucose. Free sugars in quinoa were evaluated to contain 4.55% glucose, 2.41% fructose and 2.39% sucrose. The sugar content in quinoa of Andean and India is given in Table - 6.

Table 6: *Sugar content in quinoa (g / 100 g dry matter)*

Crop	Glucose	Fructose	Sucrose	Maltose
Quinoa of Andeans	1.70	0.20	2.90	1.40
Indian Quinoa	1.68	0.23	2.91	1.42

Source: Annual Report (2015-16) of Indian Medicinal Plants Marketing Federation, New Delhi

- **Dietary fibres:** Quinoa grains are rich source of total dietary fiber (7 – 9.7% on dry basis). Dietary fibre makes quinoa an ideal food to detoxify the body, eliminating toxins and waste products that may harm the body. Fibre in quinoa is soluble and insoluble. The soluble fibre content in quinoa is reported from 1.3% to 1.6% (dry basis). Most of the fibre is insoluble

(6.8 – 8.4%) which does not appear to have the same health benefits as soluble fibre.

The total dietary fibre content in quinoa flour is affected by thermal treatment, while the insoluble dietary fibre fraction does not change with heat treatment. Boiled quinoa contains much less fibre because it absorbs so much water.

Table 7: *Carbohydrates in Quinoa, rice and barley grains (% dry basis)*

Carbohydrate	Quinoa	Rice	Barley
Carbohydrate	73.6 – 74	79.2	77.7
Starch	52.2 – 69.2	–	–
Fibre (total dietary fibre)	7 – 9.7	–	–
Insoluble fibre	6.8 – 8.4	2.8	15.6
Soluble fibre	1.3 – 1.6	–	–
Sugar	2.9	–	0.8

Carbohydrates from quinoa can be considered a nutraceutical food because they have beneficial hypoglycemic effects and induce lowering of free fatty acids.

Studies made in individuals with celiac disease showed that the glycemic index of quinoa was slightly lower than that of gluten-free pasta and bread. Besides, quinoa induced lower free fatty acid levels than gluten-free pasta and significantly lower triglyceride concentrations compared to gluten-free bread. Some nutraceutical effects of quinoa have been reported, but that requires further study.

F. FAT AND OIL

Quinoa is potential source of nutritionally rich oil and fat making, as a valuable new oil crop. Its grains contain 5.2 to 9.5% oil (average 7% on dry basis), which is similar to corn oil, but higher than corn (3 to 4.9% dry basis) and lower than soy (20.9% dry basis). The yield of extractable vegetable oil per hectare (80–400 kg/ha) in quinoa could easily exceed that obtained from maize (20–50 kg/ha). High lipid content compared with traditional cereals and essential fatty acid profile make quinoa a potential valuable oil crop.

The fat content of Quinoa oil is very valuable due to its high proportion of unsaturated fatty acids with poly-unsaturation index 3.9 – 4.7. Quinoa oil is quite stable due to high levels of natural anti-oxidants, vitamin E, 690 – 740 ppm α-tocopherol and 790 – 930 ppm γ-tocopherol. Oil of quinoa has highest%age of fatty acid i.e., 50.24 – 56% Omega-6 (linoleic acid) and 21.1 – 26.04% Omega-9 (oleic acid). Omega-3 (linolenic acid) in quinoa oil is 4.77 – 6.7% and palmitic acid accounted for 9.59%. Omega-6 and omega-3 fatty acids are essential fatty acids because they cannot be synthesized by human body, who must obtain them from foods. Additionally, a small proportion of fatty acids, such as stearic acid and eicosapentaenoic acid were also found. Quinoa oil also contains approximately 2% erucic acid. Quinoa oil is a rich source of linoleic and linolenic (essential fatty acids), which constitute approximately 55 – 63% of the oil.

In a comparison of fatty acids and triacylglycerol compositions, quinoa oil had the lowest saturate/un-saturate ratio compared with oils from Amaranthus, buckwheat, corn, rice bran, sesame, soybean and cotton seed. In addition, quinoa and soybean oils had the most favourable linoleic to linolenic acid ratio of the preceding oils.

Fatty acids have important role in:

- Signal transduction pathways.
- Cellular fuel sources.
- The composition of hormones and lipids.
- The modification of proteins.
- Energy storage within adipose tissues (specialised fat cells) in the form of triglycerols.

G. MINERALS

Minerals are chemical elements required to perform functions necessary for the life. They help our body to grow, develop and stay healthy. Human body uses minerals to perform many functions like building strong bones, transmitting in nerve impulses, make hormones and maintain normal heart beat.

Minerals are categorized in two groups i.e., essential and non-essential minerals. Essential minerals are divided into major (macro) and minor (micro or trace) minerals. These two groups of essential minerals are equally important, but, trace minerals are needed in smaller amounts as compared to major minerals. The amounts needed in the body are not an indication of their importance.

Three minerals i.e., Calcium, magnesium and iron are mostly deficient in foods and diets which contain gluten. Many modern foods are also deficient in minerals especially magnesium, potassium, zinc and iron (for women). Quinoa grains are rich in minerals like calcium, magnesium and zinc. It stands out for its high content of these elements as compared to wheat, corn, rice, barley, oats, rye and triticale. Leaves of quinoa are also rich in minerals, especially calcium, phosphorus and iron. The content of minerals in quinoa is influenced by environmental conditions and mineral availability in the soil during plant growth and seed set. Quinoa is also pretty high in oxalates, which reduce the absorption of calcium and can cause problem of certain individuals with recurring kidney stones. The mineral content in quinoa as compared to other food grains is presented in Table - 8

Table 8: *Mineral content in quinoa as compared to barley, yellow corn and wheat Grains (based on the average of 15 cultivars of quinoa)*

Crops	Ca	P	Mg	K	Na	Fe	Cu	Mn	ZN
	---------%---------				----------- ppm -----------				
Quinoa	0.19	0.47	0.26	0.87	115	205	67	128	50
Barley	0.08	0.42	0.12	0.56	200	50	8	16	15
Corn	0.07	0.36	0.14	0.39	900	21	--	--	--
Wheat	0.05	0.36	0.16	0.52	900	50	7	--	14

Source: Technical Report of Indian Medicinal Plants Marketing Federation (2017)

- **Calcium:** Calcium is an essential component of the diet which is responsible for many structural functions of hard and soft tissues of the body as well as the regulation of neuro-muscular transmission of chemical and electrical stimuli, cell secretion and blood clotting. The recommended daily intake of calcium for children (6–12 months) is 400 mg / day and 1300 mg / day for adults as per WHO. Calcium content of quinoa is between 46 to 340 mg / 100 g of dry matter. The calcium of quinoa is four times higher than that of corn, almost triple that of rice and much more than wheat. It is easily absorbable by the body, therefore, quinoa can help to prevent de-calcification and osteoporosis.
- **Iron:** Iron is essential mineral for blood production in humans and most of animals. About 70% of total iron found in the human body is present in red blood cells of blood called hemoglobin and in muscle cells called myoglobin. Hemoglobin is essential for transportation of oxygen from lungs to tissues. It is also involved in conversion of blood sugar to energy. The production of enzymes also depends on iron. Quinoa contains three times more iron than wheat and five times more than rice, while corn lacks this mineral completely.
- **Potassium:** It is one of the main blood minerals called electrolyte. The others are sodium and chlorides which carries a tiny electrical charge (potential). It is crucial to heart functions and plays a key role in skeletal and smooth muscle contraction making it important for normal digestive and muscular function. Quinoa contains potassium twice that of wheat, four times more than corn and eight times more than rice.
- **Magnesium:** Magnesium is a component and activator of many enzymes, especially those that convert energy-rich phosphates. It is also a stabilizer of nucleic acids and membranes. Quinoa contains 170 – 270 mg /100 g of dry matter. It is much higher amounts than in corn, wheat and rice.
- **Phosphorus:** The phosphorus is a mineral that is part of every cell in human body. It is found mainly in bones and teeth. It works with calcium and other nutrients to build healthy bones and theeth. Phosphorus also helps maintain normal acid / base balance, supports growth and is involved with the storage and use of energy. Phosphorus content in quinoa is similar to wheat, but much higher than rice and especially corn.
- **Zinc:** Zinc acts on the synthesis and degradation of carbohydrates, lipids, proteins and nucleic acids. Therefore, an intake between 6 and 20 mg / day is recommended. Quinoa provides 4.8 – 6.1 mg /100 g of dry matter. It is almost double the amount found in wheat and four times that of corn, while rice does not contain this mineral.
- **Manganese:** Manganese aids in the formation of connective tissue, bones, blood-clotting factors and sex hormones and plays a role in fat and carbohydrate metabolism, calcium absorption, and blood sugar regulation. It is also necessary for normal brain and nerve function. Only wheat exceeds quinoa in its manganese content, while rice has half as much and a corn only a quarter.

H. VITAMINS

An organic compound (or related set of compounds) essential for normal growth is called a vitamin. About 13 vitamins are universally recognised. Vitamins are convertible to the active form of the vitamin in the body and are sometimes inter-convertible to one another as well. The vitamin composition of quinoa is similar to that of cereals, containing significant amounts of thiamin (0.29 – 0.36%), riboflavin (0.30 – 0.32%), vitamin B-6 (0.487%) and folate (0.18%). Compared with barley, niacin values are lower in quinoa (4.604% of niacin present in barley).

Vitamin-A

Vitamin-A is important for vision, cell differentiation, embryonic development, immune response, taste, hearing, appetite and development. In quinoa seeds, Vitamin-A is in the range of 0.12 to 0.53 mg. Young tender leaves, shoots and flower heads are excellent source of vitamin-A.

Vitamin-C

Vitamin C, also known as Ascorbic acid, is needed for the growth and repair of tissues in all parts of body. It also regenerate antioxidants such as vitamin-E. Vitamin C is required for synthesis of collagen, the intercellular "Cement" substance which gives structure to muscles vascular tissues, bones, tendons and ligaments. It can be found in levels varying from 4.0 to 16.4 mg/100 g.

Viamin-E

Quinoa is excellent source of vitamin E which is present in the range of 4.60 to 5.90 mg /100 g dry matter. Vitamin E has anti-oxidant properties and prevents lipid peroxidation, thus contributing to maintaining the stability of cell membrane structures and protecting the nervous system, muscle and retina from oxidation.

Vitamin-B

Quinoa is rich source of vitamin-B complex group. Thiamine (vitamin B_1) is distributed in the pericarp of the quinoa grain and its content is in the order of 0.05 to 0.60 mg/100 g dry matter.

Vitamin content of quinoa has been presented in Table - 9

Table 9: *Content of vitamins in quinoa*

Vitamins	Range
Vitamin A (carotenes)	0.12 – 0.53
Vitamin E	4.60 – 5.90
Thiamine	0.05 – 0.60
Riboflavine	0.20 – 0.46
Niacin	0.16 – 1.60
Ascorbic acid	0.00 – 8.50

Table 10: *Macro nutrients in Quinoa and other selected foods (per 100 gm dry weight)*

Nutrients	Quinoa	Beans	Maize	Rice	Wheat
Energy(Kcal/100 gm)	399.0	367.0	408.0	372.0	392.0
Protein (g/100 gm)	16.5	28.0	10.2	7.6	14.3
Fat (g/100 gm)	6.3	1.1	4.7	2.2	2.3
Carbohydrate (g/100 gm)	69.0	61.2	81.1	81.1	78.4

Source: Annual Report (2013-14), Indian Medicinal Plants Marketing Federation,New Delhi

Table 11: *Nutritional value of cooked and uncooked quinoa per* 100 *gm*

Constituents	Un-cooked quinoa	Cooked quinoa
Energy	368 kcal	120 kcal
Carbohydrates	64.2 g	21.3 g
* Starch	52 g	48 g
* Dietary fibre	7 g	2.8 g
* Fat	6.1 g	1.92 g
* Monounsaturated	1.6 g	0.529 g
* Poly unsaturated	3.3 g	1.078 g
Protein	14.1 g	4.4 g
Vitamins		
Vitamin A	1 **μg**	0 **μg**
Thiamin (B_1)	0.36 mg	0.107 mg
Riboflavin (B_2)	0.32 mg	0.11 mg
Niacin (B_3)	1.52 mg	0.412 mg
Vitamin B_6	0.49 mg	0.123 mg
Folate (B_9)	184 **μg**	42 **μg**
Choline	70 mg	23 mg
Vitamin C	0 mg	0 mg
Vitamin E	2.4 mg	0.63 mg
Minrerals		
Calcium	47 mg	17 mg
Iron	4.6 mg	1.49 mg
Magnesium	197 mg	64 mg
Manganese	2 mg	0.631 mg
Phosphorus	457 mg	152 mg
Potassium	563 mg	172 mg
Sodium	5 mg	7 mg
Zinc	3.1 mg	1.09 mg
Other constituents		
Water	3.3 g	72 g

Source: USDA Nutrient Database - http//ndb.nal.usda.gov

Chapter 7

SECONDARY METABOLITES AND THERAPEUTIC VALUES

7.1 INTRODUCTION

Quinoa is one of the most popular and excellent "Functional and sacred foods" in the world for long and healthy life as it reduces the risk of various diseases in a preventive way. It is one of the best super foods especially for vegetarians and recognized all over the world as a delicious food grain with amazing health, skin and hair benefits. Quinoa plays a significant role in reducing blood sugar, insulin, triglyceride levels, risk of cardio-vascular disease, type-2 diabetes, high blood pressure, colon cancer and obesity. In view of these impacts, dieticians recommended to consume 48 gm of whole quinoa grains per day in 3 servings each of 16 gm to remain healthy and free from most of the diseases. Amazing health benefits and industrial values of quinoa are due to presence of secondary metabolites / bio-active compounds in its grains and leaves in sufficient proportion which have been discussed in this chapter.

7.2 SECONDARY METABOLITES

Secondary metabolites are typically organic compounds produced through the modification of primary metabolite. They do not play a role in growth, development and reproduction and are formed during the end or near stationary phase of growth. The irrigation and salinity may regulate the production of bio-active compounds in quinoa, influencing its nutritional and industrial values.

The following are the main secondary metabolites found in quinoa:

Phenolic compounds

Phenolic compounds are basically involving plant metabolic system and have potential against oxidative damages. Phenolic compounds are currently of major interest for their dietary effect. These compounds are very much essential for the growth of plant and involve in reproduction process of plants. Plants produce these compounds during response process against pathogens for defending injured plants. Due to anti-oxidant properties, phenolic compounds are widely used in processed foods as a natural anti-oxidants. These compounds have multiple properties i.e., anti-oxidant (cardio-vascular prevention), anti-allergic, anti-inflammatory, anti-viral and anti-carcinogenic. Total phenolic content in quinoa is expressed as Gallic Acid Equivalent (GAE).

Total phenolic compounds in quinoa seeds have been reported to be 71.7 mg GAE / 100 gm which is higher than that of wheat (53.1 mg GAE / 100 gm) and

Amaranths (21.2 mg GAE / 100 gm). Red seeded quinoa contain 50% more phenolic content, 90% more total flavonoid content and 150% more FRAP than yellow seeded quinoa.

Based on the number of phenol rings, phenolic compounds are divided into following two groups:

1. Simple phenols

Simple phenols include phenolic acids containing one aromatic ring with one or more hydroxyl groups.

2. Polyphenols

Polyphenols are bioactive secondary metabolites that are widely present in food stuffs of plant origin. Polyphenols contribute to the bitterness, astringency, colour, taste and oxidative stability of food products. These polyphenols act as potent anti-oxidants.

Polyphenols are of three main type:

(*i*) ***Flavonoids:*** The flavonoids are another group of phenolic compounds with 2-phenyl-1, 4-benzopyron back-bone and divided into sub-groups i.e., flavones, iso-flavones, flavan, pro-anthocyanidins and anthocyanidins. Quinoa is a more effective functional food, in terms of a source of bioactive flavonoids, than conventional cereal and pseudo-cereal grains, with a flavonoid content.

Following flavonoids are important in quinoa:

- **Quercetin and Kaempferol:** Two flavonoids i.e., quercetin and kaempferol have been particularly well studied in quinoa. They are important potent bioactive substances found in large amounts in quinoa. These two important molecules have been shown to have anti-inflammatory, anti-viral, anti-cancer and anti-depressant effects. Quinoa in diet increases intake of these important nutrients. Epi-catechin and Quercetin exhibit anti-oxidant activity and has a negative correlation with the risk of developing coronary heart disease and type-2 diabetes. Some varieties of quinoa have abundant orientin, vitexin and rutine.

  The presence of flavonoids in edible plants improves their nutraceutical value in terms of health promoting effects. Flavonoids have the potential to reduce the risk of cervical cancer, lung cancer, breast cancer and colorectal cancer. Isoflavones are a group of flavonoids that act as phyto-estrogens in animals and human beings. They also stimulate osteoprogerin secretion, improve bone health, reduce arterial resistance and exhibit anti-oxidant activity. Daidzenin and genistein are the main isoflavonoids in quinoa.

- **Anthocyanins:** It is a group of flavonoid constituents that provide the bright red orange to blue violet colours in the plants. They provide protection from DNA cleavage, estrogenic activity, enzyme inhibition, increase cytokinin production, decrease capillary permeability and fagility, and membrane strengthening. Anthocyanin in quinoa has been reported at 102.4 mg CGE/10 gm DW, higher than that of amaranth, soybean and sorghum.

- **Carotenoids:** They are anti-oxidants, presenting pro-vit-A activity, colourant and essential components of plants. Carotenoids are also reported to prevent age related macular degeneration (AMD) and benefit eye health. They can reduce the risk of coronary heart disease, protect from ischemic stroke and protect skin from UV-induced damage.

  The carotenoids in sweet quinoa is 0.4 mg / 10 gm, greater than the content in corn (0.07 mg / 10 gm), while it is about 0.08 mg / 10 mg in bitter quinoa.

(*ii*) ***Phenolic acids:*** It is a group of phenolic compounds with single phenol ring. Phenolic acids are mostly bound to cell structural components such as cellulose, protein, lignin or flavonoids and sugars by the ester, ether and acetyl linkages. Phenolic acids provide diverse functions to plant, such as nutrient uptake, enzyme actions, microbial habitation and protection against pathogens. Total amount of phenolic acids in quinoa grains varied from 16.8 to 59.7 mg / 100 g and the proportion of soluble phenolic acids varied from 7 to 61%. Gallic acid is the main phenolic acid in quinoa. *P*-Coumaric acid and Cinnamic acid are also found in quinoa with significantly higher content compared to those in whole wheat, corn and rice. The phenolic acid content in Andean quinoa is low compared to common cereals like wheat and rye, but similar to level found in oat, barley, corn and rice.

Table 12 shows that Ferulic acid in quinoa is lower than it is in whole wheat comparable to corn and higher than it is in rice.

Table 12: *Phenolic acid content in quinoa and other cereals*

S.No.	Phenolic acids	Whole			
		Quinoa	Wheat	Corn	Rice
1	Ferulic Acid	440	890	380	240
2	Caffeic Acid	40	37	26	ND
3	Vanillic Acid	43.4	15	4.6	7.8
4	Gallic Acid	320	ND	ND	ND
5	Cinnamic Acid	10	ND	ND	ND
6	*P* - Coumaric Acid	ND	37	31	76
7	*P* - Hydroxybenzoic Acid	76	7.4	5.7	15
8	Syringic Acid	ND	13	7.8	ND

(*iii*) ***Tannins:*** It is a class of antioxidant polyphenols that may impair the digestion of various nutrients. Quinoa contain tannin in the range of 0 up to 500 mg /100g.

Saponins

Uper layer of quinoa grains is called epidermis (Pricarp). This layer is distinguished as a resinous layer covering the grain. Saponins in the quinoa seed are located in the first external coat of the episperm, which is itself made up of four layers. This external coat is rough, brittle and dry, and can be partly re-moved using abrasive methods or by washing with cold water. Due to presence of saponins, quinoa grains have bitter taste and are unpalatable without processing / removal. This bitterness has beneficial effects during cultivation of quinoa, as it is unpopular with birds and therefore requires minimal protection. Removal improves considerably when warm water or alkaline or acid solutions are used.

Saponins are water soluble and form foaming solutions. The genetic control of bitterness involves quantitative inheritance; lowering the saponin content through selective breeding to produce sweeter, more palatable varieties is complicated by about 10 percent cross-pollination. Therefore, most quinoa seeds in the market have been treated to remove their cover, then washed with water or scarification, but not the sweet varieties, which have no saponins or only in concentrations of < 0.11 g /kg Dry Matter.

Quinoa varieties are classified according to their content of saponins i.e. sweet quinoa with no saponins or containing less than 0.11% and bitter quinoa. The saponin content in seeds of sweet genotypes varied from 0.02% to 0.04% and in seeds of bitter genotypes from 0.14 to 2.3%. These values are higher than those in soybean and oat, but lower than in green pea. Saponins have shown insecticidal, antibiotic, fungicidal and pharmacologic activity.

Saponins are steroid or triterpenoid glycosides compounds widely documented in quinoa. These compounds have a bitter taste and are considered toxic in large amounts. Its natural function is to defend the plant from the external medium. Saponin is mainly present in the outer layers of the grain (the pericarp) and characterized by a bitter taste and the formation of foam (saponins soluble in water), making the grain basically unpalatable. The saponin content in seeds varies between 0.2 to 0.4 in sweet and 4.7 to 11.3 g/kg dry matter (DM) in bitter genotypes of quinoa. About 20 types of triterpenoid of saponins have been described in quinoa, isolated from various parts of the plant (flowers, fruit, seed pericarp and the seeds themselves).

Saponins are the main anti-nutritional factor present in the seed cover. Studies on rats revealed that animals fed with unwashed quinoa diets showed growth damage and reduced food conversion efficiency. According to their chemical structure, saponins can be partially removed by washing with water, but even after washing some saponin remains in the seed. It is recommended to use of slightly alkaline water rather than neutral water to de-bitter quinoa. It is reported that the bitter taste imparted by saponins could potentially be reduced by extrusion and roasting processes.

Now a days, saponins have been studied to find out different beneficial properties and health benefits of saponin. Saponins possess a broad variety of biological effects i.e., analgesic, anti-inflammatory, anti-microbial, anti-oxidant, anti-viral and cytotoxic activity, effect on the absorption of minerals and vitamins and on animal growth, hemolytic and immune-stimulatory effects, increased permeability of the intestinal mucosa neuro-protective action and reduction of fat absorption. Finally, saponins have commercial–industrial importance as they are used in the preparation of soaps, detergents and shampoos. However, the biological properties of quinoa saponins require further study.

Studies of saponins isolated from the quinoa seed pericarp have revealed anti-microbial activity, toxicity in Artemia, anti-viral activity, the capacity to reduce cholesterol levels, and increased drug absorption through mucous membranes, modifying intestinal permeability. Saponins may even act as an immunological adjuvant to boost the response of gene-specific antibodies and may induce cytotoxicity and apoptosis in cancer cell lines. In addition to saponins, quinoa possesses a special class of triterpene molecules, known as phyto-ecdysteroids, which present a wide range of pharmacological effects in mammals. Ecdysteroids are steroid hormones which control molting and reproduction in arthropods. The main phyto-ecdysteroid found in quinoa plants is 20-hydroxyecdysone (20E), with an average concentration of 365 ± 51 mg/kg. Several beneficial effects on the functioning of various organs have now been recognized, for example, the anabolic property of 20E as a precursor of protein synthesis in muscle cells in humans and mice. Other studies have shown that 20E controls the regulation of blood glucose levels and activity against associated obesity. The anti-diabetic effect of this molecule has been observed in mice, through a reduction in adiposity when fat-enriched diets are used, supplemented with this phyto-ecdysteroid. Furthermore, these compounds can be considered potent chemical agents capable of preventing or retarding damage to the skin associated with the activity of the enzyme collagenase and the effects of oxidative stress.

Within the saponin of quinoa, two aglycones i.e., oleanolic acid and hederagenin predominate are found. The ratio of oleanolic acid to hederagenin varies from 2.3 to 8.6, depending upon the cultivar. Oleanolic acid and hederagenin content averaged 3.0 and 1.4 mg/g, respectively in quinoa.

At least 20 different types of saponin have been identified in quinoa. These chemical compounds include various monosaccharide units that are attached via a glycosidic bond to a triterpene skeleton, known as an aglycone or sapogenin. Depending on the number of saccharide chains in the structure, they may be classified as mono, di or tridemosidic. Monodesmosidic saponins contain a single saccharide chain, generally located in C-3. Bidesmosidic saponins contain two saccharide chains, one of them generally attached by an ether bond to C-3, and the other attached to C-18 or C-26 by an ester bond. The most common monosaccharides are D-glucose, D-galactose, D-glucuronic acid, D-galacturonic acid, L-rhamnose, L-arabinose, D-xylose and D-fructose. Four aglycones have been identified in quinoa saponins: oleonolic acid, phytolaccagenic acid, hederagenin. Some authors count serjanic acid as the fourth aglycone, while others consider it to be spergulagenic acid.

The physico-chemical and biological properties of saponins have been used in many commercial applications in the food, cosmetics, agricultural and pharmaceutical sectors. Despite being considered as anti-nutritional substance like tannins, phytic acid and protease inhibitors and although it has a negative effect on red blood cell levels in blood types A and O, there is scientific proof of their beneficial health effects due to their anti-carcinogenic properties and cholesterol lowering effect. Some studies have also demonstrated their anti-fungal properties.

In current characterizations, various quinoa varieties and ecotypes are designated as "bitter", "semi-sweet" and "sweet". This classification is based on saponin content, which is generally 0.3% in dry grains. Saponin content in "bitter" grains is 1–3%, in "sweet" grains - 0.0–0.1% and in "semi-sweet" grains 0.1–1%. Other quinoa scientists believe that a variety or ecotype may be considered "sweet" if the saponin content is

20– 40 mg/100 g dry weight, and "bitter" if the saponin content is > 470 mg/100 g dry weight. The maximum acceptable limit of saponin content in the cooked grain is 0.1%.

Subsequently, the rest of the episperm and the abrasives are winnowed away from the grain for 20–40 minutes. In the final stage of saponin removal and of removal of impurities, such as small stones and seeds collected during harvest, the quinoa grains are washed in various stages over 25–35 minutes. The effluent is inspected visually to check for foam formation, the quality control parameter. When the effluent is clear of foam, this indicates that the saponins have been removed. Finally, the quinoa is dried for 2–4 hours, until the final moisture content is about 18%. Depending on what it is to be used for, the grain may sometimes be winnowed and roasted again.

Other bioactive components

- **Squalene:** It is a highly unsaturated open-chain triterpene, which is the biochemical precursor of the whole family of steroids. About 3.39 - 5.84% of squalene is present in the lipid fraction of quinoa seed. This level of squalene is higher if compared to other plant oils such as olive oil, wheat germ oil and rice bran oil i.e. from 0.1 to 0.7%. As a food constituent, squalene inhibits the ability of cholesterol synthesis in the liver by lowering the cholesterol levels.

7.3 THERAPEUTIC VALUES

A. Health benefits:

- **Aphrodisiac benefit:** Quinoa plays an important role as aphrodisiac in the problem of male potency, which refers to the ability to have better erection in males and reduced infertility in women. The term potency generally refers to the ability to have an erection and hence have the power, energy, strength and capacity to engage in sexual intercourse. Very often, when human males are concerned about the problem of male potency, they describe it as a problem of erectile dysfunction. When females are investigated for the problem of infertility, the investigation of the male partner sometimes reveals the problems such as very low sperm counts and other abnormalities associated with the seminal fluid. For men, impotence involves the problem of getting sexually excited, the stimulation of sexual desire for the successful execution of sexual intercourse.

  The recommendation of quinoa as an aphrodisiac is based upon other considerations. Androgens are sex hormones that influence the growth and development of the male reproductive system. Testosterone is produced by the male testes. In the male, the interstitial cells of Leydig in the connective tissue surrounding the sperm-producing tubules of the testes are responsible for the production and secretion of androgens which are essential for the growth of the reproductive organs, the formation of sperm cells and for the maintenance of sexual interest and libido or sexual drive. Secondly, Polyamines are required to protect the sperm cells during their journey in the female genital tract. Polyamines are alkaline compounds that counteract and buffer the acidic environment of the vaginal canal and protect DNA inside sperm cells. From a study of Phytochemistry, plant physiology, human physiology, biochemistry, medical biology, endocrinology, embryology and human nutrition, it could be recommended to include some dietary sources of polyamines during the reproductive activity of both men and women to ensure the birth of viable and healthy babies.

Studies indicate that reducing the concentration of Polyamines in the body pool may slow cancer process as the rapidly proliferating cancer cells need polyamines. To study viruses, Virologists use *Chenopodium quinoa* plants as plant virus hosts. The development of Polyamine Database for assessing Dietary Intake has been recommended and we are hopeful that nutritional guidelines will become available. Quinoa is a functional food and it must be used in support of various physiological functions and its intake and of other polyamines could be also important in the management of cancer. The biosynthesis of chemical molecules by plants is a creative process and the mechanism is creative as it generates diversity, while the molecules maintain their unchanging properties and characteristics.

- **Anti-inflammatory:** The fibre of quinoa produces butyrate, an important fatty acid, that turns off genes related to inflammation. Vitamin-B of quinoa reduces homocysteine levels (an inflammatory hormone) in the body. The saponin present in grains have anti-inflammatory properties.
- **Healing properties:** The grain of quinoa contains alkaline substances (calcium, magnesium and trace elements) that are used as a natural remedy for sprains strains by means of making a plaster of black quinoa mixed with alcohol and applying it to the affected areas to heal broken bones.
- **Anti-gastric properties:** Due to its content of starch and soluble fibre, quinoa is a demulcent grain which protects mucous membranes or the skin inside the organs. So, it can help ease sore stomach. Decoction is effective in gastritis and stomach pain.
- **Benefit in atherosclerosis:** Quinoa is rich in quercetin compound which is beneficial in reducing atherosclerosis (hardening of arteries), circulatory problems and coronary heart diseases.
- **Role in reducing high blood pressure:** Quinoa is Fibre rich food which reduces the risk of high blood pressure and heart attack. High insoluble fibre helps to reduce the secretion of excessive bile acids.
- **Prevent osteoporosis:** People have more concern about the health of their bones. The individuals above fifty years must be really concerned about bone health. Quinoa is rich in magnesium and Manganese which work great for bone formation and maintaing bone health. Protein of quinoa also serves as a building block for the bones.
- **Anti-oxidant properties:** Quinoa is loaded with anti-oxidants (flavone glycosides) and it can serve as a good source of free radical scavenging agents. Anti-oxidants are substances which neutralize free radicals in the body

and are believed to help fight aging and many diseases. Quinoa has higher anti-oxidant activity than some cereals (rice and buck wheat). Allowing the seeds of quinoa to sprout seems to increase the anti-oxidant content even further.

- **Helpful in Gonorrhea and Tuberculosis:** Due to its high mineral content, quinoa is used as an anti-blenorrhagic remedy for gonorrhea and tuberculosis. The infusion of quinoa leaves is used as a laxative for the treatment of constipation.
- **Beneficial in migraine:** Quinoa is good source of magnesium, manganese, folate and phosphorus that are helpful for migraine headaches. The magnesium and vitamin B-2 in quinoa help to relax blood vessels and improve the energy metabolism, thus, reducing the frequency of headache episode.
- **Helps in prevention of gallstone:** High content of insoluble dietary fibre in quinoa can reduce the risk of gallstone by lowering blood triglycerides and reducing the stagnation of bile, that aid gallstone formation.
- **Helps in weight loss:** Quinoa is a great satiating food having control on weight and obesity. First, it is zero-cholesterol, low fat grain. Second, it is rich in dietary fibres that help to keep the feeling of satiety, thus, reducing hunger pangs and cravings. Quinoa grains has less single sugar and more complex carbohydrates, more healthy fats and other vitamins and minerals necessary for weight loss. Quinoa offers a perfect weight loss solution as it is low in terms of calories. Presence of 20-hydroxyecdysone compound in quinoa grains is known to help with weight control due to its high potential of burning more calories which consequently aids weight control. It also makes individuals absorb less fat from their diet.

  In order to lose weight, we need to take in fewer calories than we burn. It is known that certain properties of foods can facilitate this process, either by boosting metabolism (increasing calories out) or reducing appetite (lowering calories in).
- **Safe for diabetes:** Quinoa has low glycemic index i.e., 53 and glycemic index is a measure of how quickly foods raise blood sugar levels. The foods having high glycemic index can stimulate sugar level and contribute obesity. Since quinoa have low glycemic index, it is a safe food for diabetic patients. The fibre of quinoa prevent raising of sugar level in blood, diabetes related weight gain and other chronic conditions. The complex carbohydrates of quinoa breakdown in the body much slowly, allowing the blood sugar to be more stable.

  Some research findings are contradictory and reveal that quinoa has no alpha-amylase inhibitory activity that is necessary to control diabetes. Manganese function as a co-factor for the saferoxide dismutase enzyme - an anti-oxidant. Some studies show that soluble fibre can help to reduce blood sugar level, lower cholesterol, increase fullness and help with weight loss.
- **Protects heart and Improve Cardio-vascular health:** Quinoa binds cholesterol and fat in the body and reduce the building of plaque in the arterial

walls of heart. As a result, it reduces the risk of heart strokes and other heart ailments. Studies have shown that deficiency of magnesium increases the angiotensin II that is responsible for increasing blood pressure. Magnesium rich in quinoa relaxes blood vessels and prevent high blood pressure. Quinoa contains fatty acids, most of them are in the form of oleic acid which keeps heart healthy.

- **Helps fight cancer:** As per study of Harvard University, cancer is beatable and preventable by eating a bowl of quinoa daily. It reduce the risk of premature death by cancer. The extract of quinoa leaves exert chemopreventive and anti-cancerous effects. Quinoa is super rich in anti-oxidants that can fight free radicles and other harmful substances that cause cancer.

B. Skin benefits

- **Aids in quick healing of damaged tissues:** Quinoa contains lysine, a key building block in the synthesis of elastin and collagen. This aid in quick healing of damaged tissue, making it very important for people who are recovering from any surgery or injury. Lysine is an essential nutrient which cannot be synthesized by the body, hence it must be provided by the diet. Quinoa also helps to create scar tissue and ligaments for quick recovery.
- **Provide elasticity and builds connective tissue:** Quinoa contains high level of Vitamin-2 (riboflavin) which provides elasticity to the skin to give it the required softness and resiliency. It also builds up connective tissues to tone the skin and prevents wrinkles and fine lines.
- **Anti-ageing face pack:** Quinoa contains natural protein and anti-ageing properties, while soy milk nourishes and repair sun damaged skin to give elasticity to the skin. Cooked quinoa blended with soy milk and yoghurt act as anti-ageing face pack for glowing skin. This face pack also make the skin smooth and even toned.
- **Rejuvenate the skin:** Quinoa grains provide the body with ample collagen which rejuvenates the skin from within and reduces wrinkles and other symptoms of ageing. It also makes the skin soft, smooth and supple. Vitamin-A present in quinoa delay the aging process. It reduces fine lines and makes look of skin young.
- **Reduces brown spots and pigmentation:** B-Vitamins and nutrients present in quinoa help to treat age spots and other conditions related to pigmentation by reducing the deposits of dark melanin granules in the skin. Vitamin B-12 of quinoa interacts with the other B-Vitamins to maintain a healthy skin complexion. Quinoa also contain tyrosinse inhibitor enzyme that decrease pigmentation and associated problems.
- **Treats acne:** Vitamin B-2 and 3 of Quinoa helps to decrease sebum production and brilliantly treats acne. It soothes the red and inflamed areas often associated with acne breakouts.

C. Hair benefits

- **Protects and nourishes hair follicles:** Hydrolyzed protein extracted from quinoa acts as a natural and gentle coating which protects and nourishes the hair follicles from within. The protein extracted from quinoa grains is also used to produce high quality hair products.
- **Protects hair shaft:** Nine essential amino acids present in quinoa act as natural strengtheners and protect the shaft of the hair. The amino acids of quinoa also repairs damaged hairs and promote hair growth.
- **Promote hair growth:** Regular consumption of quinoa provides sheen to hair and gives it an overall silky and shiny look. It provides nourishment to the cells of the body including our hair follicles, which is essential for our hair growth. It is also essential for a healthy nervous system and for the formation of red blood cells which promotes hair growth.
- **Treats dandruff:** The minerals like calcium, iron and phosphorus present in quinoa grains seal moisture in the scalp to avoid flaky and dry scalp, and keep dandruff at bay.
- **Nourishment and hydration to scalp:** The humectants present in quinoa provide nourishment and hydration to the scalp to keep it well conditioned. It forms an invisible film over the hair to protect it from environmental conditions like pollution.
- **Reduce breakage:** Vitamin E of Quinoa balances the production of natural oil to keep the strands properly moisturized and reduces breakage due to tangling.
- **Maintain hair colour:** Tyrosine amino acid present in quinoa is responsible for the pigmentation and colour of hair and skin. It encourages the re-pigmentation of hair and even helps to maintain the original colour of the hair.

D. Other benefits

Quinoa is considered suitable for the treatment of cholesterol in diets, for hypertension and vascular disease prevention in general. Quinoa also improve glucose tolerance, reduces insulin requirement and reduces blood cholesterol. It is also recommended especially for contribution of dietary fibre for people who are operated on the gall bladder and diabetes.

In addition, the amino acid proline, contributes to joint repair. Proline is not found in other grains like wheat. It is necessary for healing wounds and ulcers, appears to be effective in treating cases of impotence and frigidity, is a cardiovascular protector and is used together with lysine and vitamin C to prevent or limit cancer metastasis. Glutamic acid is involved in the production of energy for the brain and important phenomena, such as learning, memory and neuronal plasticity. Cysteine protects the liver by binding to heavy metals and facilitating their removal in addition to destroying free radicals and boosting the immune system. Serine is potent natural moisturizer and tyrosine has an important anti-stress effect and plays a fundamental role in alleviating depression and anxiety, among other functions.

According to traditional medicine, the quinoa's stems and leaves cooked with oil, vinegar increase the formation of blood. The leaves cooked with vinegar and used as a gargle or applied as a poultice to heal sore throats and anginas. A decoction of the leaves prepared with sugar and cinnamon purifies the stomach, dislodges phlegm and bile and removes nausea and heart burn. An infusion of the leaves is used to treat urinary tract infections or as a laxative. The fresh leaves of quinoa 'chiwa', consumed either as a soup or dessert, are a remedy against scurvy and other illnesses or diseases caused by vitamin deficiency.

Quinoa fruit contains a large amount of alkaline substances and is used as a remedy for sprains, fractures and dislocations, making a paste by mixing it with alcohol or brandy. A decoction or poultice made from the fruits is used medicinally to apply on wounds and bruises. The water from the cooked grain cures liver abscesses, internal secretions and catarrhal afflictions. Likewise, the cooking water from the cooked grain mixed with milk and almond oil is used to wash the ears where there is pain, noise and deafness.

Good sudorific is produced by cooking five table spoons of quinoa seeds in two bottles of water. The same decoction, sweetened with honey or molasses, is a proven remedy against bronchial disorders, colds, cough and inflammation of the tonsils. The broth, soup, or warm grain of quinoa is nutritive tonic, increases breast milk, is restorative and protects against tuberculosis. Quinoa soup immediately increases the milk supply of lactating women.

The protein in the Quinoa oil press cake could be an important complementary protein for improving the nutritional quality of both human and animal food-stuffs. Little work has been reported exploring the technical and commercial feasibility of these ideas.

Chapter
8

ECOLOGICAL REQUIREMENTS

8.1 INTRODUCTION

The life-cycle of quinoa plants vary from 120 to 240 days depending upon the climatic conditions and nutritional availability. Quinoa is well suited to various environmental conditions as it has ability to adapt to adverse environmental conditions, such as cold and drought. Quinoa, in general, is un-demanding, altitude-hardy and drought-resistant plant which remarkably adapted to different agro-ecological zones. It is mainly grown from coastal regions to over 4000 metres in the Andes (South America) near the equator. Most of the cultivars of quinoa are grown in between 2500 metres and 4000 metres altitude. Depending upon the variety, quinoa crop matures normally in 90 to 125 days after seeding. A major constraint for growth in northern parts of Europe, Canada and in high altitude regions is the short growth season, because Quinoa requires a longer growing season, in order to secure seed harvest. Therefore, early maturing cultivars are recommended for the areas where growing season is relatively short.

The following phenological phases have been recognised in quinoa:

- Emergence stage; 2, 4 and 6 true leaves,
- Branching stage;
- Ear formation stage;
- Start of inflorescence;
- Florescence or anthesis;
- Woody grain; soft grain
- Physiological maturity.

8.2 CLIMATE REQUIREMENT

Temperature, light (photoperiod), rainfall, humidity, frost and their seasonal variations relate to climate factors. Quinoa requires mainly cool climate for its cultivation; however, it adapts from desert to hot /dry regions. The climatic requirement of quinoa is described here under:

- **Temperature:** Temperature plays an important role in establishment of quinoa plants. It affects germination, growth and reproduction of the plants. Quinoa can grow at temperatures between 15° to 20°C and can withstand temperatures from - 8°C during the night to 38°C during the day. Quinoa has been cultivated in the high elevation where maximum summer temperatures rarely exceed 30°C and night temperatures are about 7°C. Extreme hot weather and warm nights inhibit fruit set. Some varieties of quinoa tolerate lower temperatures upto - 4°C in the branching and grain forming stages without any damage, depending on the ecotype and duration of minimum temperature. Ecotypes exist which are resistant to temperature of down to - 8°C and survive for 20 days (mean monthly temperature). Temperature sensitivity is highest in cultivars originating in cold or dry climates and lowest for cultivars from humid and warmer climates. Cold tolerance depends on the stage of development when the frost occurs and the natural protection of the mountains. Its ontogenic resistance to cold and drought is very variable.

Air temperatures, which exceeded 35°C tended to cause plant dormancy or pollen sterility. Quinoa plants are usually tolerant to frosts (–1.1 to 0° C).

Plants exposed to below –2.2°C at mid bloom stage could reduce seed yield upto 70% due to flower abortion. However, quinoa can tolerate the temperature as low as – 6.7°C after the grain has reached the soft-dough stage.

- **Photoperiod:** Photoperiod response on quinoa plants governs the flower and seed production. Light is a prime factor controlling growth and life of quinoa plants. After seed germination, all developmental phases of quinoa plants still remain under control of light. Light hastens growth and maturation of young coleoptile cells, because those cells begin to elongate earlier and mature earlier than in darkness. The radiation sensitivity in quinoa is reported to be highest in cultivars from Peru, Bolivia and Southern Chile and lowest in those from Ecuador. The experiments conducted by the author of this book in controlled environments under project entitled "Evaluation of production potential of quinoa in marginal areas of Indian Sub-continents", have revealed that most of cultivars of quinoa taken from Andean region had a facultative short-day response for duration of emergence to flowering and the duration of all the phases of development is sensitive to photoperiod. The studies on the effect of radiation on quinoa further revealed that short-day treatment showed a quantitatively positive response on the anthesis and total number of leaves, while maximum seed growth was obtained under short day and cool temperature treatment. Another study has shown that photoperiod sensitivity of phyllochron decreased as the latitude of origin of the cultivar increased. Quinoa requires short day length for early flowering and cool temperatures for excellent growth.
- **Rainfall:** Quinoa is a highly tolerant and resistant plant to lack of soil moisture. It can produce acceptable yields in regions where the annual rainfall is in the range of 100 – 200 mm. The growth of quinoa plants is reported to optimal with well-distributed rainfall during early growth and development stage. The optimum growth of quinoa plants was also recorded under dry conditions during seed maturation and harvesting.
- **Humidity:** Quinoa plants are grown best at relative humidities ranging from 40 to 88 percent. Extreme hot weather and warm nights inhibit fruit set.
- **Frost:** Frost is one of the principal limiting factors for crop production in the areas dominated by severe frost. Quinoa is quite hardy and frost resistant

plant, generally not affected by frost as compared to other crops, but, little is known about its specific mechanisms for frost resistance. Studies conducted by the author of this book under project entitled "Evaluation of productiion potential of quinoa in marginal areas of Indian Sub-continents", revealed that frost causes certain damage of quinoa plants during flowering stage resulting reduction in yield. Light frost normally does not affect the plants at any stage of development especially if the soil is dry. It was further concluded that frost causes significant damage when it occurs during flowering as flowers are sensitive to frost. When green stage of seed maturing passed, frost will cause little damage. Frost if occurs during flowering stage, it may lead to sterilization of the pollen. If the stage of seed maturing is finished, frost will cause little damage.

8.3 EDAPHIC FACTORS

Among edaphic factors, soil (fertility, pH, soil temperature, radiation) and water influence growth and development of quinoa plants. Their role and effects on growth and development of quinoa plants are discussed below:

- **Soil requirement:** Quinoa can be successfully grown in poor soils, but heavy clay soils are not suitable for its successful production. The most suitable soils for quinoa cultivation are sandy to sandy loam. Quinoa perform best in semi-deep, well drained and nutrient rich sandy soils with moderate salinity. Soil acidity and alkalinity have considerable influence on growth and development of quinoa plants. It tolerates a wide range of acidic conditions and salinity with the soil pH 4.5 to 9.5 depending on ecotype. Soil pH affects the availability of plant nutrients. Therefore, pH value of 7 is very much suitable for better growth of quiona plants. A very high or very low soil pH may result in plant nutrient deficiency or toxicity. The microbial activity and nutrient absorption from the soil by the plants is reported to be the highest at pH from 6 to 7.5.

  Quinoa is reported to be one of the few crop plants grown in the salt-affected soils. However, high salinity influences plant growth, seed yield and seed quality of quinoa. Plant growth, total seed yield, number of seeds, fresh weight and dry weight of seeds per plant are reduced in the presence of salinity. At high salinity, protein content increases in seeds, while total carbohydrate content decreases. Organic matter is needed to improve soil structure, slow release of nutrients and increase in the activity of beneficial micro-organisms.

  Soil acidity is also a major growth limiting factor for quinoa plants worldover. Aluminium toxicity is a major problem under acidic soils. Low pH in top soils primarily affects nutrient availability and ultimately decreases the growth of quinoa plants. Aluminium is present in soils in a variety of forms and bound to the soil constituents, particulary clay particles and organic matter. When pH drops, aluminium becomes soluble and its amount increases in soils. When, its concentration in soils exceedes 2.5 ppm, it becomes toxic to the roots of sensitive plant species. Due to its high concentration, root tips of quinoa and all other are deformed and brittle affecting growth and branching. Quinoa plants tolerate its concentration in soil upto 5 ppm. In acidic soils, if organic manures applied in desired amount, quinoa can be grown easily upto 5 soil pH as sufficient organic matter in top soil remain bound and does not become toxic to the roots of quinoa plants. Liming soils to increase the soil pH is effective in reducing the availability of aluminium to non-toxic level.

Water requirement

Water is crucial to all life. Although, Quinoa is most drought hardy plant, however, it needs water for growth and survival. There are three potential situations with water i.e., too much, too little and, of course, just enough. If soil has too much water, the roots can rot and the plant can't get enough oxygen from the soil. If there is not enough water for quinoa plants, the nutrients it needs cannot travel through the plant. Water helps a plant by transporting essential nutrients drawn from the soil for use of plants. Without enough water in the cells, the plants droop. So the proper balance of water is key when quinoa is being grown.

Quinoa is a water efficient crop with low water requirement and is tolerant and resistant to the lack of soil moisture. The low water requirement shows its drought-hardy nature and makes it suitable for cultivation on large tracts of India where assured irrigation is non-existent and farmers have to depend on seasonal rains. 2 to 3 irrigations are recommended for obtaining good yields of quinoa. Excess watering or more irrigations in the seedling stage causes diseases like stunting and damping off, while such conditions after stand establishment produce tall plants with no yield improvement.

Chapter 9

SCIENTIFIC PRACTICES OF CULTIVATION

9.1 FIELD PREPARATION

The seedbed must be well prepared for obtaining better yield of quinoa. Two to three ploughings followed by planking are needed for field preparation. There should be a better drainage system to avoid water logging condition.

If the soil is clay or coarse sand (not beach sand), compost and peat moss can be added. In sandy soils, humus or well-decomposed manure, peat moss or saw-dust can be added with some extra nitrogen. Heavy clay-rich soil can also be added to improve the soil. Well decomposed manure is good soil conditioner, hence, it should be added in silty soils, coarse sand (not beach sand) and gravel to improve soil texture.

Well decomposed Farm Yard Manure (FYM), good quality compost or well-rotten horse manure can be mixed in the soil with fresh straw. Decomposed leaf mold may be added in the soil to raise nutrient status and to improve structure of the soil. In acidic soils, lime can be added in the soil to raise soil pH and helps loosen clay soil. Peat moss is considered as best soil conditioner which may be added to improve water retaing capacity of the soil.

Fig. 9. *Planking for field levelling*

Fig.10: *Field preparation for quinoa sowing*

9.2 VARIETIES

The varieties of quinoa brought from Andean Region and other countries are not successfully adopted and suitable in different Indian agro-climatic conditions. There is no research work conducted on Quinoa in India and availability of improved varieties are mostly lacking. Inspite of urgent need of quinoa in food and health securities, varietal availability is the main hindrance in achievement of quinoa revolution in India.

The author of this book has developed a high yielding variety of quinoa under project entitled "Evaluation of productiion potential of quinoa in marginal areas of Indian Sub-continents", i.e., Sarvati (S-7). The special features of this variety have been discussed here under.

SARVATI (S-7)

Only one variety of quinoa i.e., Sarvati (S-7) has been developed and available so far in India for quinoa cultivation. It has been developed by the author of this book Dr. Dharm Singh, a pioneer Agriculture Scientist in India. It was developed during 2010 and released in 2015 by Indian Medicinal Plants Marketing Federation for scientific cultivation in almost all parts of Indian sub-continent. About 13 genotypes were brought during 2002 from Andean Region of Peru and Chile, and tested under project entitled "Evaluation of production potential of quinoa in marginal areas of Indian Sub-continents", being implemented by Indian Medicinal Plants Marketing Federation. Two varieties i.e., **Villarica** genotype of Chile and **Ragalona** genotype from Peru performed well in different regions in India under testing and their yield was recorded highest (5 – 6 Q / Acre) among all genotypes. These varieties contain highest protein (16.5%), but, both were found low yielding and susceptible to grey mold and bacterial blight diseases at different places under Indian conditions.

Their performance was recorded satisfactory at the temperature up to 35°C. After doing 5 years breeding work on these two genotypes, an improved and high yielding variety was developed, named "Sarvati (S-7)".

Special features: The special features of Sarvati (S-7) variety of quinoa are given below:

- It is highly suitable for growing under different agro-climatic conditions of Indian sub-continents.
- It yielded from 8 to 19.5 Q Acre^{-1} at farmer's field under Front-line Demonstrations.
- It contains highest protein (16 to 19%) depending upon farming conditions.
- It is heat tolerant and grown successfully at the temperature up to 38° Celsius.
- Its plants are multi-branched which bear economic panicles up to 16.
- It is highly resistant to grey mold (*Botrytis cinere*a) and bacterial blight (*Pseudomonas species*) diseases of quinoa under Indian conditions.
- It is sweeter and palatable as its grain contain very less saponin in concentration less than 1.2 g Kg^{-1} dry matter.
- It is recommended for growing throughout India under irrigated and marginal environment.

Sangamner (Maharashtra)

Faridabad (Haryana)

Fig. 11 (a): *Panicle size of sarvati* (5-7)

Allahabad (Uttar Pradesh)

Fig. 11 (b): *Size of panicle of Sarvati (S-7) variety at different locations in India*

Fig. 12: *Grains of Sarvati (S-7) Variety*

9.3 SOWING

Seed sowing is one of the most important activities in quinoa production because the emergence of seedlings, which imparts plant density and final yields, depends on this activity. A levelled and well-drained seedbed is generally needed for seed sowing. The time of seed sowing, method of sowing, climatic conditions, type of varieties and soil moisture are important factors which determine the plant population in the field and ultimate yield of quinoa grains. These are also important factors in determining the sowing method to be used i.e., manual or mechanical. The method and time of sowing of quinoa are described below:

Time of Sowing:

The time of seed sowing of quinoa in India varies from August to October in the high lands. In the Southern states, the best time of quinoa seed sowing is from late August to mid-November, while in the Central and Northern regions, the planting time from mid October to mid November is found suitable. When soil temperatures at the time of quinoa seed sowing are around 15°C, seedlings emerge within three to four days. However, when quinoa seeds are planted in soil with night-time temperature is much above that, the seeds of quinoa may not germinate. In such conditions, it is recommended to refrigerate seeds of quinoa for a week before planting. Quinoa is a warm season crop that requires full sun. Best germination occurs when soil temperature range from 18–24°C at the time of sowing.

Method of sowing:

In the Andes, the seeds are normally broadcasted over the land and raked into the soil. Sometimes, seeds are sown in containers and seedlings are transplanted later. In India, seeds may be sown by broadcasting method without maintainig proper distance among lines. In this method, weed control and mechanized practices become difficult in the field. Therefore, line sowing is recommended for easy cultural operations and to maintain optimum plant stand in the field. Seed sowing manually in lines is found excellent for geting higher yield. Seeds of quinoa may be sown in rows at spacing range between 40 to 50 cm. Seeds should be sown when moisture is sufficient in the soil. Sowing by seed drill is not possible in India as maintaining recommended seed depth in the soil is very defficult. This sowing system is efficient on uniform level ground.

Fig. 13: *Quinoa seed sown in lines on raised beds.*

The plant density in quinoa may vary according to the region. Around 20 to 30 plants per square metre are the most appropriate. To maintain this densiy of plants, 6 to 8 kg seed / ha should be maintained.

Fig. 14: *Quinoa seed sown by broadcasting*

9.4 WEEDING

Weed control has major impact on grain yield of quinoa. Weeds pose a serious threat to quinoa during early stage upto first two weeks after emergence because of its slow growth. During this stage, weed competition from rapidly growing weeds is greater.

An early sowing would enable quinoa to have a head start over weeds as the plant can attain good growth during this period. Herbicides are not recommended in the cultivation of quinoa. This is more important to control weed population in quinoa by hand weeding to maintain quality of grains.

9.5 IRRIGATION

Quinoa has low water requirement due to its drought hardy nature. Plants should not be irrigated until the two or three leaf-stage. Excess irrigation after establishment usually produce tall, lanky plants with no yield improvement. Damping off and severe stunting of plants are noticed under testing with excessive irrigation in the seedling stages. Watering in quinoa by drip system is found excellent at different places in India under Front-Line Demonstrations conducted in different states under project entitled "Evaluation of production potential of quinoa in marginal areas of Indian Sub-continents".

Fig. 15: *Drip irrigation in quinoa at Sangamner (Maharastra)*

Fig. 16: *Irrigation in quinoa through furrows at Akole (Maharastra)*

Light irrigation in furrows can also be given in quinoa for better results (Fig. 16).

9.6 HARVESTING:

- **Time of harvesting:** The maturity time of quinoa crop depends on many factors i.e., variety, soil type, humidity and prevailing temperature. Quinoa is harvested at the stage of physiological maturity. When the crop is physiologically matured, it should be harvested timely without any delay.

The best way to determine the optimum stage of harvest is to gently but briskly shake or rub the flower heads between your hands and see if the seeds fall readily. Quinoa crop generally get matured after 90 days and its harvesting stage comes when the leaves have fallen and leaving just the dried seed-heads. At this stage, the grains become dry and hard. Numerous small and appreciative birds may give hints as to when to start harvesting.

Fig. 17: *Heavy bearing crop (Variety- Sarvati) at Sangamner (Maharashtra)*

Fig. 18: *Bumper crop of quinoa (Sarvati-S 7) at Faridabad (Haryana)*

Fig. 19: *Bumper crop of quinoa (Sarvati-S 7) visited by Scientists/officers of Govt. Departments at Sangamner (Maharashtra)*

Rain and clousy weather at the time of harvesting, for more than one day without existence of bright sun light, are very dangerous. Under this situation, quinoa seeds starts germinating on the plants. Longer the cloudy and rainy weather, greater will be the germination. It is important to watch the weather when quinoa is ready for harvesting.

Fig. 20: *Quinoa bumper crop (Variety-Sarvati) at Hyderabad (A.P.)*

- **Method of harvesting:** If the heads are not completely dry, harvest them when you can barely indent the seeds with your thumb nail. Seeds can be easily stripped upwards off the stalk with a gloved hand. Usually there are three ways to harvest the plants:
- **Traditional uprooting:** Traditional method of quinoa harvesting is uprooting of the plants. This method consists in pulling out the plants by hand, selecting the mature panicles from each hole or row and proceeding to shake or hit the root on the knees in order to reduce the presence of earth clods and pebbles. The disadvantage of this method is that it does not leave the root in the soil as organic matter and also contributes to soil erosion, lowering soil fertility. This method also leads to soil being mixed with the grain, increasing the presence of impurities during threshing.

Fig. 21: *Bumper crop of quinoa (Variety-Sarvati) at Sangamner (Maharashtra)*

- **Harvesting with a sickle:** Manual harvesting with a sickle consists in cutting the plants between 10 and 15 cm above the soil and leaving the stubble in the soil, thus, helping soil conservation. This method may be followed at the time of proper maturity. The plants must be cut at the right time, i.e., when the panicles still resist the loss of grain upon handling, because when the plants exceed maturity there are increased grain losses. The disadvantage of this method is that it cannot be used on very sandy soils and it is difficult to cut large plants because of the thickness of the stem.
- **Semi-mechanical harvest:** Semi-mechanical harvesting involves cutting the plants with a mechanical mower. Its use is made easier when the plants are arranged in holes or furrows. If the heads are not completely dry, harvest them when you can barely indent the seeds with your thumb nail.

9.7 THRESHING AND CLEANING

Quinoa can be threshed mannualy or by small threshers. Grains may be cleaned by using screens, winnowing with a fan or other blowing device.

Fig. 22: *Mannual threshing of quinoa*

9.8 YIELD

Generally, 7 to 8 quintal grain yield per Acre is obtained under normal conditions. If quinoa is grown scientifically, 8 to 15 quintal grain yield can be obtained. The author of this book (Dr.Dharm Singh) has obtained quinoa yield upto 19.5 quintal per Acre under front-line demonstrations .

9.9 STORAGE

Quinoa grains have long self-life. It can be stored safe for longer duration. It should not be rotten or spoiled during storage. For safe grain storage, quinoa grains should be stored in a cool and dry place in air-tight container with tight fitting lids. The store should be away from sunlight and heat. A properly sealed container is vey important for maintaining the freshness and reducing the possibility of infestation. In this way, quinoa grains stay fresh for several months or for over a year and not become rancid or smelly.

Chapter

10

SEED PRODUCTION TECHNOLOGY

10.1 INTRODUCTION

Seed is a basic input and foundation of crop production. Seed quality plays an important role in succesful production of crops. Without using high quality seeds, yield and crop quality would be greatly decreased. Seed is an embryo, a living organism embedded in the supporting or the food storage tissue. In seed, the importance is given to the biological existance where as in grain, the importance is given to the supporting tissue the economic produce. Production of high quality quinoa seed requires considerable technical skill and a number of rigid requirements must be fulfilled in order to ensure a high purity and high germination of seeds.

As per seed act, 1966, a mature ovule, consisting of an embryonic plant together with a store of food and surrounded by a protective coat. Usually developed upon fertilization by male and female gamete. The different important issues related to seed production of quinoa have been discussed in this chapter.

10.2 FOOD RESERVES IN QUINOA SEEDS

There are following three food reserves in quinoa seeds:

(i) A large central perisperm.

(ii) Peripheral embryo.

(iii) One to 2 cell layered endosperm.

Carbohydrate / starch reserves are present principally the perisperm, while protein, mineral nutrients and lipid reserves are located in the endosperm and embryo. The perisperm consisted of uniform, non-living, thin walled cells in angular shape. The cells are full of starch grains. The Embryo is peripheral and basal body is present in the seed as a storage tissue or perisperm. All embryo cell stored abundant protein and lipids in the form of protein and lipid bodies. Protein bodies contained one or more globoid crystals in the proteineous matrix and varied in diameter.

In the mature seed, the endosperm is present only in the micropylar region of the seed and consists of one or two cell-layered tissue surrounding the hypocotyle-radicle axis of the embryo. The endosperm cells had thick hard outer-cell walls. The starch of grains occupy the cells of the perisperm, while lipid bodies, protein bodies with globoid crystals of phytin and proplastids with deposits of phytoferritin are the storage components of the cells of endosperm and embryo tissues in the micropylar region developing the hypocotyl-radicle axis.

10.3 STRUCTURE OF QUINOA SEED

The seeds of quinoa are generally flat, oval-shaped which are produced in large clusters on a panicle. Seed size and colour are variable. The quinoa seeds measure about 2 mm in diameter. Seeds can be black, red, pink, orange, yellow or white in colour depending upon the cultivar type. The seed colour is due to resinous coating that contain 2 to 6 percent saponins. Black seeds are dominant over red and yellow, which in turn are dominant to white seed colour.

The quinoa seeds are commposed of three basic structures:

- **The seed covering (seed coat or testa):** Episperm surrounding the grain is composed of four layers. The outer layer which determines the colour of the seed is rough, brittle, easily removed with water and contains bitter substances known as saponins, which protects grains from birds due to toxic and bitter taste. It is necessary to eliminate saponins from the seeds before consumption / eating or manufacturing of food products. Seed covered by epidermis represent 60 percent part of the seed. It is most rich in starch. Epidermis is developed by the embryo from which yelloyish white tail like structure, called radical, stands out.
- **The embryonic axis:** The embryonic axis normally includes the miniature plant, consisting of the root (radicle) and shoot (Plumule). The radicle in quinoa seed constitutes 30 percent of total seed volume wrapping with a curature of 320 degree.
- **Supporting tissues:** Supporting tissues in quinoa seeds usually include the cotyledons and endsosperm. They are useful for the developing plant as a reserve food source through course of germination and emergence until the plant can make its own food through photosynthesis. Supporting tissues in quinoa are composed mostly of non-living, starchy materials, where as such tissues in dicots are composed of mostly fats and oils.

In quinoa, the main storage reserves is clearly divided in 3 main storage components which can be distinguished in the mature seeds as described below:

- A tissue that stores mainly starch, which is the perisperm in quinoa.
- An embryo that store principally protein and lipids which is the richest part of the seed in proteins (35 – 40%).
- Non-embryonic tissue that stores proteins, lipids and minerals which is the aleurone layer in gramineae and the micropylar cone in quinoa.

The embryo comprises 60 percent of volume within the pericarp and these results in the higher protein content of the seed in comparison to cereal grains. This embryo is the richest part of the seed in protein (35 to 40%).

The seed coat consisted of two layers of cells i.e., exotesta and the endotegmen. The cells of exotesta are large with very thick walls and reduced cytoplasm, starch grains and crystals are occasionly observed in the cytoplasm of these cells. The cells of endotegmen showed small thickening in the inner walls.

10.4 CATEGORY / TYPE OF SEEDS

There are following five categories of seed:

- **Nucleus seed:** This is 100 percent genetically pure seed which produced by the original Breeder / institute / State Agriculture University from basic nucleus seed stock. A pedigree certificate is issued by the producing breeder.
- **Breeder seed:** The progeny of nucleus seed multiplied in large area as per indent of Department of Agriculture, Ministry of Agriculture, Government of India, under supervision of plant breeder / Research Institutes / State Agriculture Universities and monitored by a committee consisting of the representatives of the state seed certification Agencies, State / National Seed Corporation, ICAR nominee and concerned breeder. It is 100 percent genetically and physically pure seed used mainly for production of foundation seeds.
- **Foundation seed:** The progeny of breeder seed produced by recognised seed producing agencies in public and private sector, under supervision of seed certification agencies in such a way that its quality is maintained according to prescribed field and seed standartds. A white coloured certificate is issued and used by seed certification agencies for this category of seed.
- **Certified seed:** The progeny of foundation seed produced by registered seed growers under supervision of seed certification agencies to maintain the seed quality as per minimum seed certification standards. Certified seed is produced under strict varietal purity standards and identified with a blue label. The basic purpose of production and sale of certified seed is to make genetically pure seed available to the farmers. A blue coloured certificate is issued by seed certificaton agencies to the producer of seed of this category.

  The foundation and certified seeds can be multiplied at stage-I and stage-II, but the reproduction cannot exceed three generations.
- **Truthfull Lebel (TL) seed:** It is category of self-certified seed. Truthful labelling is compulsory for notified kind of varieties and quality is guaranteed by producing agencies, applicable to notified and released varieties, tested for physical purity and germination.

10.5 SEED QUALITY PARAMETERS

Seed quality is the possession of seed with required genetic and physical purity that is accompanied with physiological soundness and health status. The major seed quality parameters are discussed hereunder:

- **Physical purity:** It is the cleanliness of seed from other seeds, debris, inert matter, diseased seeds and insect damaged seeds. The seed with physical quality should have uniform size, weight and colour and should be free from stones, debris and dust, leaves, twigs, stems, flowers, fruits, well without other crop seeds and inert material. It also should be devoid of shriveled, diseased mottled, moulded, discoloured, damaged and empty seeds. The lack of this quality parameters will indirectly influence the field establishment and planting value of seed. This quality character could be obtained with seed lots by proper cleaning and grading / processing of seed after collection and before sowing / storage.

- **Genetic purity:** It is true to type nature of the seed, i.e., the seedling / plant / tree from the seed should resemble its mother in all aspects. This quality character is important for achieving the desired goal of raising the crop either yield or for resistance or for desired quality factors.
- **Physiological quality:** Physiological quality of seed include potential of seed viability, germination and vigour.
  - **Seed viability:** The liveliness of a seed or potential of seed to germinate is known as seed viability. All viable seeds need not be germinable, but all germinable seed need to be viable. A non-viable seed will not germinate under any condition. The state of non-germination in viable seed is known as dormancy. The germinaion is blocked by conditions within the seed.

    Dormant seeds are often in resting state. Seed dormancy is broken simply by passage of time, removal or breaking of seed coverings, cold over-wintering or the effects of light or hormones supplied to the seeds. Requirements of seed germination include a favourable moisture level in the seed, temperature in environment around the seed and oxygen supply to the seed.
  - **Seed germination:** The extent of liveliness for production of good seedlings or the ability of seed for production of seedling with normal root and shoot under favourable condition is known as germinability. Quinoa seeds do not exhibit dormancy and its seeds have a notably short germination period. When adequate moisture is present, germination occurs within 24 hours after planting and seedlings emerge in 3 to 5 days. Quinoa seeds may not germinate if conditions are warm and may need to be refrigerated for a week to vernalize them to obtained adequate germination. Seeds may remain viable in the soil for two or three years without germination and germinate immediately when conditions are suitable.
  - **Seed vigour:** Seed vigour is the energy or stamina of seed that gives seed the potential for rapid and uniform emergence and development of normal seedlings under a wide range of field conditions. Seeds vigour determines the level of performanc of seed or seed lot during germination and seedling emergence. All vigourous seeds will be germinable, but all germinable seed need not be vigourous. Hence, selection of seed based on seed vigour is an important for raising perfect finalize plantation.
- **Seed health:** Seed health is nothing but absence of insect infestation and fungal infection, in or on the seed. Seed should not be infected with fungi or with insect-pests as these will reduce the physiological quality of the seed and also the physical quality of the seed in long term storage. The health status of seed influences the seed quality characters directly and warrants their soundness in seed for the production of elite seedlings at nursery / field.

Hence, the quality seed should have :

- High genetic purity.
- High pure seed (Physical purity) percentage.
- High germinability.
- High vigour.
- Higher field establishment.
- Free from insect-pest infestation.
- Good shape, size, colour etc. according to the specification of variety.
- High longevity / shelf life.
- Optimum moisture content for storage.
- High market value.

10.6 METHOD OF QUINOA SEED PRODUCTION

- **Land requirement:** The field to be used for seed production should not have been used for growing the same crop in the previous year. If the same crop was grown in previous year, the field should be irrigated 3 weeks before sowing and ploughed just before sowing. This is done to allow the seeds that might have dropped in the field from the previous crop to germinate and to destroy the seedlings by ploughings. In addition, the selected field should be well-drained and the soil deep, fertile and with neutral pH.
- **Isolation requirement:** The quinoa crop raised for seed production should be separated from other fields of the same crop species by a mimum distance of 5 meters. This distance is known as Isolation distance. Isolation is essential to prevent pollination from unwanted pollen in case of cross pollinated species and to avoid mechanical mixture and chance of cross pollination in self-pollinated species.

  Quinoa is partially self-pollinated. The extent of cross-pollination varies from 5 to 10 percent according to insect activity. The field of quinoa seed crop should, therefore, be isolated at least by 400 meter in case of foundation seed and 200 meter for certified seed production from fields of other varieties / same variety not conforming to varietal purity requirements and wild quinoa.
- **Cultural practices:** Seed production requires improved cultural practices, efficient management of weeds, diseases and pest control, optimum irrigation and fetlizer inputs and some of other specific operations. Field preparation, method of seed sowing, irrigation, weeding, harvesting and other recommended cultural operations should be followed as explained in previous Chapter-9 and plant protection measures as explained in chapter-14 & 15.

Specific weeding operations:

- **Removal of objectionable weeds:** Effective weed control is essential for good quality seed production. Weeds reduce crop yields and also contaminate the seeds. Certain weeds are classified as objectionable weeds and rogues by the seed certification agency and the field of seed crop is required to be free from such weeds. If any objectionable weed and rogue are found at the time of field inspection, the seed field may be rejected.

- **Roguing:** The plants of other varieties of same crop in seed crop under certification are known as off-type or rogue and removal of rogue is known as *roguing*. It is important aspect to prevent out crossing and mechanical mixture. Generally two roguings are necessary in quinoa seed production. The first should be done at pre-flowering stage and the second at crop maturity. Before flowering, tall, very early, very late, branched as well as weak, wild and diseased plants should be rogued out. At the time of 75% crop maturity, wild, ornamental and diseased, damaged and all those plants which do not conform to the characteristics of the variety under seed production should be rogued out. In addition to these, plants affected by wilt, any type of rot, blight, etc. should also be removed from time to time as required. The off-type / rogue plants may differ in plant height, leaf characters, flowering time, maturity etc. The off-type / rogue should not be allowed to mature and be removed regularly from the field either by uprooting or by cutting at the ground level.

Fig. 23: *Mannual weeding and roguing in quinoa seed crop*

10.8 STANDARDS FOR QUINOA SEED

- Genetic purity :
  - Breeder / Nucleous seed – 100%
  - Foundation seeds (Minimum) – 99.5%
  - Certified seed (Minimum) – 99%
- Physical purity (Minimum) – 99%
- Germination (Minimum) – 70%
- Other crop seed (Number / Kg) – 20
- Inert matter (Number / Kg) – 20
- Moisture content (Maximum) – 6 – 8%

Chapter

11

FERTILIZER MANAGEMENT

11.1. INTRODUCTION

Plants normallly require nutrition from the soil, water and air for their growth and development. The production and quality of crop produce (grains) depend on balanced feeding of crops. Therefore, for better growth of the plants, high economic yield and excellent quality of grains, fertilizers should be applied in proper quantity at proper time. Fertilizer management is, therefore, an important aspect in cultivation of any food crop with a view to get good quality higher grain yield. Scientific fertilizer management in quinoa with other related necessary issues have been discussed in this chapter for better understanding.

11.2 ESSENTIAL NUTRIENTS REQUIRED

Analysis of a plant reveals the presence of a large number of mineral elements which may differ from plant to plant, place to place and medium to medium in which the plants grow. But all these elements found in a plant are not essential for its growth and life cycle. A large number of them are non-essential. Only 16 elements are considered essential for every botanical plant including quinoa which are listed in Table-13 alongwith their sources.

Table 13: *Essential plant nutrients and their source*

<table>
<tr><th>S.No.</th><th>Nutrients</th><th colspan="2">Category</th><th>Source</th></tr>
<tr><td>1</td><td>Carbon (C)</td><td colspan="2" rowspan="3">Structural elements
(which constitute 94 to 99.5% body of the plants)</td><td rowspan="2">From Air</td></tr>
<tr><td>2</td><td>Oxygen (O)</td></tr>
<tr><td>3</td><td>Hydrogen (H)</td><td>From water</td></tr>
<tr><td>4</td><td>Nitrogen (N)</td><td rowspan="3">Primary nutrients, also called fertilizer elements</td><td rowspan="6">Macro-nutrients or Major elements</td><td rowspan="13">From soil</td></tr>
<tr><td>5</td><td>Phosphorus (P)</td></tr>
<tr><td>6</td><td>Potassium (K)</td></tr>
<tr><td>7</td><td>Calcium (Ca)</td><td rowspan="3">Secondary nutrients</td></tr>
<tr><td>8</td><td>Magnesium (Mn)</td></tr>
<tr><td>9</td><td>Sulphur (S)</td></tr>
<tr><td>10</td><td>Iron (Fe)</td><td colspan="2" rowspan="7">Micro-nutrients. Also called Trace elements or Minor elements</td></tr>
<tr><td>11</td><td>Manganese (Mg)</td></tr>
<tr><td>12</td><td>Boron (Bo)</td></tr>
<tr><td>13</td><td>Zink (Zn)</td></tr>
<tr><td>14</td><td>Copper (Cu)</td></tr>
<tr><td>15</td><td>Molybdenum (Mo)</td></tr>
<tr><td>16</td><td>Chlorine (Cl)</td></tr>
</table>

Carbon and Hydrogen constitute about 95 percent or more of the weight of plant. Crop plants obtain most of their carbon and oxygen directly from the air by photosynthesis. The hydrogen is derived, either directly or indirectly from the soil water. The remaining 13 nutrients (4 to 16 from table-13) are obtained by plants from the soil with the exception of leguminous plants which can fix and utilize a considerable amount of Nitrogen from the atmosphere through bacteria found in root nodules.

Six nutrients i.e., Nitrogen (N), Phosphorus (P), Potassium (K), Calcium (Ca), Magnesium (Mn) and Sulphur (S) are used by plants in relatively large amounts, they are designated as macro or major nutrients. Out of these 6 macro-nutrients, fertilizer manufacturers and also every plant gave first preference to Nitrogen (N), Phosphorus (P) and Potassium (K), thus, these are designated as primary nutrients. Since most of the soils of the world are deficient in these three elements, they are supplied through the application of fertilizers. The fertilizer industry has been producing and supplying these nutrients to the farmers in large quantities in the form of commercial fertilizers and hence, they are also called fertilizer elements. Due to secondary importance to the manufacturers out of Nitrogen (N), Phosphorus (P), Potassium (K), remaining 3 nutrients i.e., Calcium (Ca), Magnesium (Mn) and Sulphur are called secondary nutrients.

Out of 16, seven nutrients i.e., Iron (Fe), Manganese (Mg), Boron (Bo), Copper (Cu), Molybdenum (Mo) and Chlorine (Cl) are used by field crops in very small quantities, mostly less than 1 ppm, however, they are essential just as macro-nutrients. These 7 nutrients are called micro-nutrients. Other synonyms given to these nutrients are trace, minor or rarer elements.

11.3 FUNCTIONAL CLASSIFICATION OF ESSENTIAL PLANT NUTRIENTS

As seen from the Table 13, certain plant nutrients are constituent of plant tissues, while some nutrients act as catalyst and regulate various physiological and metabolic activities during growth and development of the plants. Accordingly, the plant nutrients may be classified on the functional basis in following four groups.

Group I: These are not minerals in origin, but, they present in plant and account for a major part of the dry weight.

S.No.	Nutrient	
1	Carbon	Constitutes about 45% of the total tissues of plants.
2	Hydrogen	Constitutes about 43% of the total tissues of plants.
3	Oxygen	Constitutes about 6% of the total tissues of plants.

Group II: This group contains accessory structural elements of the more active and vital living tissues, particularly the proteins. Figures are given in percentage in total tissues.

S.No.	Nutrient	
1	Nitrogen	1 to 3%. Constitutes 16 to 18% of all proteins.
2	Phosphorus	0.05 to 1 %. Constituent of nucleo-proteins, phytin and lecithin.
3	Sulphur	0.05 to 1.5%. Constituent of many proteins, oils, and often present as sulphate.

Group III: Regulators and carriers for the most part (figures given in percentage range in total tissue)

S.No.	Nutrient	
1	Potassium	0.3 to 6%. Present as salts and not constituents of plant tissues, functions mainly in synthesis and translocation of carbohydrates.
2	Calcium	0.1 to 4.0%. Present as calcium pectate and proteinat and as simple organic and inorganic salts.
3	Magnesium	0.05 to 1.0%. Present as essential constituents of chlorophyll and may function as a carrier of phosphorus. Calcium and magnesium also serves as constituent of some plant compounds.

Group-IV: Catalyst and activators for the most part (figures given in range contents in parts per million of total tissue)

S.No.	Nutrient	
1	Iron	10 to 1000 ppm (parts per million). Play role in formation of chlorophyll and probably in oxidation-reduction reactions; also exists in organic combinations.
2	Manganese	5 to 500 ppm. Believed to function in chlorophyll sysnthesis and oxidation-reduction reactions.
3	Zinc	5 to 100 ppm. Probably functions as catalytic agent.
4	Copper	2 to 50 ppm. Probably functions as catalytic agent.
5	Boron	2 to 75 ppm. Function in growth process is not known, might be function as calatytic agent.
6	Molybdenum	0.01 to 10 ppm. Only very minute amounts needed probably for catalytic purposes.
7	Chlorine	100 to 250 ppm. Required by plants in greatest amount among all micro-nutri-ents. In fact, the normal chlorine content of certain plants is as large as most macro-nutrients. Thus, its position seems to be intermediate between macro and micro-nutrients.

For proper growth and development of quinoa plants, nutrients should fulfil the following conditions:

- Nutrients must be present in the soil in available form for easy absorption by plants. For example, phosphorus is absorbed by plants as PO_4, HPO_4 or H_2PO_4.

In general, the HPO_4 and H_2PO_4 ions are some what more available to the plants than the PO_4 ions. In soil, phosphorus may be present as tricalcium phosphate $Ca_3(PO_4)_2$. For becoming more useful to the plants, it must, therefore, change into water soluble or mono-calcium phosphate $Ca(H_2PO_4)_2$ form.

- The nutrients must be present in optimum concentration for optimum growth of quinoa plants. Deficiency or excess of any nutrient usually causes limiting or harmfull effects on plant growth. This condition is particularly important for micro-nutrients which are required in small amounts.
- There must be a proper balance among the concentrations of the various soluble nutrients in the soil solution. For example, too much calcium may interfere with phosphorus and boron nutrition and may favour yellowing of plants due to reduction in the availability of iron, zinc or manganese in the soil.

11.4 UTILIZABLE FORMS OF PLANT NUTRIENTS

Plants absorb the nutrients from the soil in following two ways:

(a) As single nutrient or in un-combined with other elements.

(b) As nutrients combined with other elements to form nitrate, phosphate, sulphate etc.

Nutrients absorbed in un-combined forms:

Nutrients	Forms
Potassium	K^+
Calcium	Ca^{++}
Magnesium	Mg^{++}
Iron	Fe^{++} (Ferrous), Fe^{+++} (Ferric)
Manganese	Mn^{++} Manganous), Mn^{+++} (Manganic)
Copper	Cu^+ (Cuprous), Cu^{++} (Cupric)
Zinc	Zn^{++}
Chlorine	Cl^-

Nutrients absorbed in combined forms / as salt:

Nutrients	Forms
Nitrogen	NH_4^+ (Ammonium), NO_3^- (Nitrate)
Phosphorus	PO_4(Phosphate), HPO_4(Phosphate), H_2PO_4(Phosphate),
Sulphur	SO_3^- (Sulphate), SO_4 (Sulphate)
Boron	BO_3 (Borate), HB_4O_7 (Biborate)
Molybdenum	HM_0O_4 (Molybdate)
Carbon	CO_3 (Carbonate), HCO_3 (Bicarbonate)
Water	H^+, OH^-

Thus, mineral nutrients become available to the plants both in ionic and molecular forms, the proportions depending on the nature of the soil solutions and the conditions. Most of the nutrient uptake by plants is in the form of ions. When compounds or salts in solution exist as electically charged portions called ions. Also the ions penetrate the absorbing tissues of plants with greater ease than the molecular forms or compounds or salts. Hence, molecular absorption is trifling. The only exception of major importance is the rapid absorption of urea molecules ($CO\ NH_2\ NH_2$) through the leaf tissues and other above ground vegetative portions.

11.5 FUNCTIONS OF ESSENTIAL PLANT NUTRIENTS

The functions of various essential nutrients in physiology and development of quinoa plants are given below:

Nitrogen

It is an essential constituent of proteins, nucleic acids and many other organic molecules, such as chlorophyll. Nitrogen is present in many other compounds of great physiological importance in plant metabolism, such as nucleotides, phosphatides, alkaloids, enzymes, hormones, vitamins, etc. Thus, it is basic element of plant life.

The main functions of Nitrogen are given below:

- Imparts dark green colour to the plants.
- Promotes leaf, stem and other vegetative growth, but retains small root system.
- Improves quality and succulence of leaves and increases protein content in leaves and grains.
- Governs utilization of potassium, phosphorus and other elements to considerable degree.
- Produces rapid early growth.

Phosphorus: Phosphate is found about 0.2 to 0.8 percent of the total dry weight of plant. It is present abundantly in the growing and storage organs of the plants, such as fruits and seeds.

The main functions of phosphorus are as under:

- Phosphorus is a constituent of nucleic acid, phytin and phospho-lipids. An adequate supply of phosphorus early in plant life is important in laying down the primordial for the reproductive parts of the plants.
- Phosphorus is also constituent of majority of enzymes which are of great importance in the transformation of energy, in carbohydrate metabolism, in fat metabolism and also in respiration (Catabolism of carbohydrates) in plants. It is closely related to cell division and cells development.
- Stimulates early root development and growth, thereby helping to establish seedlings quickly.
- Gives rapid and vigorous start to plants, strengthens straw and decreases lodging tendency of quinoa plants.
- Brings about early maturity of plants particularly the cereals including quinoa and counteracts the effects of excessive nitrogen.

- Stimulates flowering and aids in seed formation.
- Increases ratio of grains to straw.
- Improves quality of the food-grains and other crop produce.
- Excess of phosphorus may cause in some cases trace element deficiencies particularly iron and zinc, but may at times alleviate the detrimental effects of over-liming.

Potassium

Unlike all other major nutrients, potassium does not enter into the composition of any of the important plant constituents, such as proteins, chlorophyll, fats and carbohydrates. As such, its role is deficult to determine. It occurs in a state of solution in the cell sap. Being soluble, it can be removed with solution in water from the plant tissues.

The main functions of potassium in quinoa plants are as under:

- Imparts increased vigour and disease resistance to the plants.
- Produces strong, stiff straw in cereals, including quinoa, paddy and wheat. Thus, it reduces lodging in cereals.
- Imparts hardiness to the plants.
- Regulates water condition within the plant cells and water loss from the plant by maintaining the balance among anabolism, respiration and transpiration. It thus, reduces tendency to wilt and helps in better utilization of available water.
- It is essential in the formation and transfer of starches and sugars.
- Helps in formation of proteins and chlorophyll.
- Increases plumpness of grains and seeds.
- Acts as an accelator of enzymes action i.e., diastase.
- Counteracts the injurious effects of excess nitrogen in plants. Hence, a balance ratio of nitrogen and potassium is important in plant nutrition of quinoa.
- Improves the quality of quinoa grains with improvement in protein, fibre content and taste, size and keeping quality.

Calcium

- Calcium is the chief constituent of middle lamella of the cell wall. As such, it increases stiffness of straw.
- It promotes early root development and growth.
- It provides a base for neutralization of organic acids, commonly termed as poison produced in the plant.
- It is essential to activate growing points, especially root tips and it does not move freely from the older to the younger parts of plants. This is the main reason why calcium deficiency symptoms are first manifested at the tips of shoots and growing points of quinoa plants.
- It influences the water economy of the plant, the proteins–carbohydrate ratio in fat metabolism as well as many other physiological process.

The effects of calcium, however, are antagonistic to those of potassium. Therefore, an optimum ratio of potassium is of great importance for the favourable water balance.

- It improves intake of other plant nutrients, specially nitrogen and trace elements such as iron, boron, zinc, copper and manganese by correcting soil pH.
- It encourages seed production.
- It helps in translocation of carbohydrates.

Magnesium

Magnesium is easily leached and for this reason may become deficient in sandy soils during wet periods. Heavy application of potassium fertilizers reduces its absorption. It is readily mobile and when its deficiency occurs, it is apparently transferred older to younger tissues where it can be re-utilized in growth process. As a result, its deficiency symptoms develop first on older leaves.

The main functions of magnesium in the plants are as under:

- It is a constituent of chlorophyll and, therefore, essential for all green plants for formation of this pigment. It helps in maintaining dark green colour in leaves.
- It plays a part in the production of carbohydrates, proteins, fats and vitamins and certain catalytic reactions in the enzyme systems.
- It acts as a carrier of phosphorus in plants, particularly in connection with the formation of seeds of high oil contents and fats which contain compound lecithin.
- It helps the translocation of starches and sugars.
- It regulates the uptake of other nutrients and the base economy of plants.

Sulphur

Sulphur is available to plants from the soil in the form of soluble sulphates. The characteristic odour of cruciferous plants, onion and garlic is due to sulphur as a constituent of volatile oils. Sulphur is known for its un-even distribution in various organs of a plant. Its main functions in quinoa plants are as under:

- Though, sulphur is not a constituent of chlorophyll, it helps in chlorophyll formation and encourages vegetative growth of the plants.
- It is an essential constituent of amino acids (cysteine and methionine) of quinoa proteins, vitamin-B1, co-enzyme A and certain volatile compounds of oil and helps in reduction-oxidation system in respiration and in the activation of ferments.
- It increases root development of plants.
- Stimulates seed formation.

Boron

Boron occurs in rocks and marine sediments. Main functions of Boron are as under:

- The primary role of boron appears to be concerned with calcium metabolism, both with its uptake by roots and its efficient use in plants.
- It tends to keep calcium soluble and increases its mobility in the plants.
- It acts as a regulator of potassium ratio in the plants.
- Boron helps in absorption of nitrogen.
- It is constituent of cell membrane and is essential for cell division. In boron deficiency, cell division ceases at the growing point, while the individual cells continue growth.
- Hence, the deficiency sysmptoms of boron, like those of calcium, are more marked at the growing tips of shoots and roots.
- It is concered with precipitation excess cation, buffur action, maintenance of conducting tissues and with regulatory effect on other elements.
- It is necessary for translocation of sugars in the plants.
- It helps in the vascular system in roots to give out rootlets to supply nodule bacteria with carbohydrate-food so that bacteria may not become parasitic.
- It has effect on many other functions in plant such as active salt absorption, hormone movement, flowering and fruiting processes, pollen germination, carbohydrate as well as nitrogen metabolism, metabolism of pectic substances, respiration, water metabolism and water relations in the plants.
- It is necessary for translocation of sugars and is involved in the reproduction and germination of pollens.
- It is concerned with water reactions in cells and regulates the intake of water into the cell.

Manganese

It is available from the soil in oxide forms, but the more highly oxidized forms (manganese oxide) are of very low availability to plants. Its solubility increases with the increased acidity and in strongly acid soils. It is frequently present in toxic concentrations. Its main functions in plants are as under:

- It increases the solubility of iron by oxidation, hence, in certain cases abundance of manganese leads to iron deficiency. It also supports the movement of iron in the plant.
- It acts as an activator of some oxidases, peroxidases, dehydrogenases, kinases, decarboxylases etc and is essential for the formation of chlorophyll.
- It acts as catalyst in oxidation and reduction, reactions within the plant tissues. Thus, as a constituent of enzymes, it helps in respiration and in protein sysnthesis in the chloroplast.
- A good manganese supply some times helps in counteracting the bad effect of poor aeration.

Iron

It is fairly present in the form of its oxides giving red or brown colour to the soil. In well irrigated areas, ferric compounds are predominantly found and in water-logged soils, ferrous compounds are formed. The availability of iron to the plants increases with acidity and is depressed by phosphates. It is absorbed in ferric state, but metabolically it is active in ferrous state. Its main functions in the plants are as under:

- Though, it is not a constituent of chlorophyll, it helps in its formation. A deficiency of iron causes chlorosis among veins of leaves.
- Iron is a very immobile element within the plant, as such; iron deficiency is noticeable in younger leaves at the growing region.
- It helps in absorption of other nutrients.
- As a constituent of enzymes, it regulates respiration, photosynthesis, reduction of nitrates and sulphates. These reactions are essential to the plants development and reproduction.
- It is essential for sysnthesis of proteins contained in the chloroplast.
- Iron is a constituent of cytochromes, ferredoxin, catalase, peroxidase, fluorochrome, hematin, globoids of aleurone grains etc.
- It also acts as an activator of nitrate reductase and aconitase.

Zinc

It is found in soils in very small quantities and largely results from concentration and addition from growing plants and added residues. In the neighbourhood of zink deposits, it is generally found to be toxic. Its uptake is reduced due to large or prolonged supply of phosphate fertilizers. Its main functions in plants are as under:

- Zinc is a constituent of several enzymes like carbonic anhydrogenase, alcohol dehydrogenase, lactic dehydrogenase, glutamic dehydrogenase, alkaline phosphatase, carboxypeptidase etc which regulate various metabolic reactions in the plant, for example, oxidation-reduction reactions in the formation of chlorophyll.
- It influences the formation of some growth hormones in the plants.
- It is helpful in reproduction of certain plants.
- It is associated with water uptake and water relations in the plants.
- It has been found essential for carbon di-oxide evolution and utilization, carbohydrate and phosphorus metabolisms and synthesis of RNA and auxins.

Molybdenum

It is found in soils and plants in very small amounts, but its higher concentration occurs in mineral oils and coal ashes. Its main functions in the plants are as under:

- It acts in enzyme system which bring about oxidation-reduction reactions, especially in reduction of nitrated to ammonia prior to amino acid and protein systhesis in the cells of the plant.

- It is essential for the formation of nodules in leguminous crops and process of atmospheric nitrogen fixation, both symbiotic and non-symbiotic.

Copper

Copper is found in the soils in small quantities resulting from additions from the growing plants and added residue. Soils neighbouring the copper deposits are normaly toxic to plants. Factors affecting its availability to plants are little known, but organic matter, soil organism and pH are the important.

The main functions of copper in the quinoa plants are as under:

- Copper forms many compounds with amino acids and proteins in the plants.
- It acts as electron carrier in enzymes which brings about oxidation-reduction reactions in plants.
- It helps in utilization of iron in chlorophyll synthesis.
- When applied in relatively large quantity, it neutralizes the harmful conditions in certain peat soils by precipitating or inactivating certain toxins present in the soil.

Chlorine

The exact role of chlorine in nutrition of quinoa plants has not yet been clearly defined. The need of chlorine for proper plant development has been established for sugarbeets, carrots, cabbage, lettuce, barley, wheat, cotton and clovers.

11.6 NUTRITIONAL DEFICIENCY SYMPTOMS

Nitrogen (N)

Symptoms of N-deficiency in quinoa plants are general chlorosis of lower leaves, stunted and slow growth, and necrosis of older leaves in severe cases. N deficient plants will mature early and crop quality and yield are often reduced. In most cereals including quinoa, yellow discolouration from the leaf tip backward in the form of a 'V' is common. Insufficient amounts of N in cereals will result in few tillers, slender stalks, short heads and grain with low protein content. Fields deficient in N can be either uniform or patchy in appearance, depending on the cause of the deficiency. For example, lack of fertilizer application will result in uniform strips of deficiency, whereas N deficiency as a result of soil characteristics such as organic matter content will be patchy.

In absence or low supply of nitrogen, the main Nitrogen deficiency symptoms in quinoa and other cereals are noticed as under:

- Stunted growth of plants due to reduction in cell division and protein, and lateral buds remain dormant. As a result, cereals do not show tillerings.
- Pale green to light yellow colour (chlorosis) appearing first on older leaves, usually starting at the tips. Depending on the severity of deficiency, the chlorosis could result in the death and/or dropping of the older leaves.

This is caused by the translocation of Nitrogen from the older to the younger tissues.

- Reduced Nitrogen lowers the protein content of quinoa and other cereal grains / seeds and vegetative parts. In severe cases, flowering is greatly reduced.
- Nitrogen deficiency causes early maturity in some cereal crops, which results in a significant reduction in yield and quality (protein and starch content).
- The plants always look thin and poorly furnished.
- Flowering and fruitings are reduced.
- Shoots remain short and thin with upright spindly growth.
- Prolonged dormancy and early senescence appear.

Phosphorus (P)

Plants require P for the development of ATP (energy), sugars and nucleic acids. Cool soils during the early growing season may be a factor causing phosphorus deficiency. P deficiency symptoms are usually more noticeable in young plants, which have a greater relative demand for P than more mature plants. Older leaves are affected first and may acquire a purplish dis-colouration due to the accumulation of sugars which favours anthocyanin synthesis; in some cases, leaf tips will brown and die. From a field perspective, phosphorus deficiency is likely to occur on tops of ridges or other exposed areas that have highly eroded or weathered soils, or in areas that have been leveled, exposing sub-soils high in calcium carbonate ($CaCO_3$). Crops grown in soils high in $CaCO_3$ are prone to P deficiency due to the precipitation of insoluble Ca-P minerals.

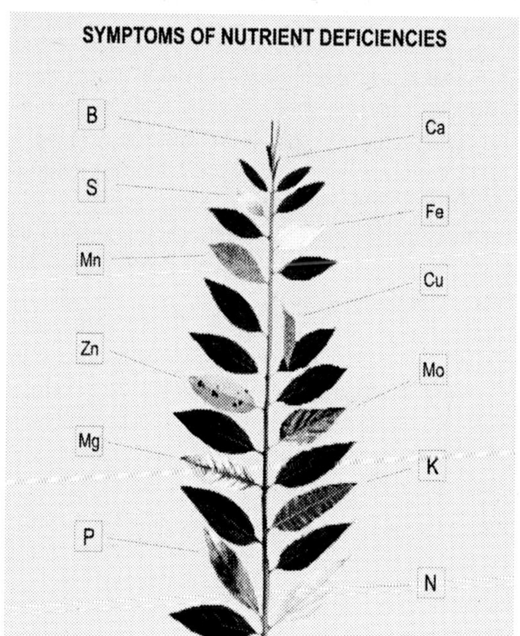

Fig. 24: *Nutritional deficiency symptoms*

The main phosphorus deficiency symptoms in cereals including quinoa are noticed as under:

- P deficient plants generally turn dark green (both leaves and stems) and appear stunted.
- Plants suffering from P deficiency appear weak and maturity is delayed.
- Leaf expansion and leaf surface area may also be inhibited, causing leaves to curl and be small.
- Small grains with P deficiency tend to be stressed and pre-disposed to root rot diseases and some cultivars will turn red or purple.
- Since P is needed in large quantities during the early stages of cell division, the initial overall symptom is slow, weak and stunted growth.
- P is relatively mobile in plants and can be transferred to sites of new growth, causing symptoms of dark to blue-green colouration to appear on older leaves of some plants. Under severe deficiency, purpling of leaves and stems may appear.
- Lack of P can cause delayed maturity and poor seed and fruit development.
- Leaves sometimes develop anthocyanins in veins and may become necrotic.
- Tillering of crop plants is reduced.
- The dormancy is prolonged.
- Thickening of tracheidial cell is reduced and phloem differentiation is found incomplete.

Potassium (K)

Potassium is utilized by plants in the activation of enzymes, photosynthesis, protein formation and sugar transport. K deficiency does not immediately result in visible symptoms (hidden hunger). Initially, there is only a reduction in growth rate, with chlorosis and necrosis occurring in later stages. Affected older leaves will show localized mottled or chlorotic areas with leaf burn at margins. Chlorotic symptoms typically begin on the leaf tip, but unlike the 'V' effect caused by N deficiency, K deficient chlorosis will advance along the leaf margins towards the base, usually leaving the midrib alive and green. As the deficiency progresses, the entire leaf turn yellow. Small white or yellow necrotic spots may also develop, beginning along leaf margins. Another indication of K deficiency is reduced straw or stalk strength in small grains and corn, resulting in lodging problems, reduced disease resistance and reduced winter-hardiness of perennial or winter annual crops. Some small grains, especially barley, develop excessive numbers of tillers when K is deficient. Grain will be low in protein and appear shriveled. In alfalfa, white spots will appear on leaf edges. Due to K's role of potassium in sugar accumulation, root crops (i.e., potatoes, sugar beets) will have small tubers.

The most common potassium deficiency symptoms in cereals and quinoa are as under:

- Chlorosis occurs along the edges of leaves (leaf margin scorching) in interveinal region. This occurs first in older leaves, because K is very mobile in the plant. The colour of the leaf may be dull or bluish green.

- Since K is needed in photosynthesis and the synthesis of proteins, plants lacking K will have slow and stunted growth.
- In some cereal crops, stems are weak and lodging is common if K is deficient.
- The size of seeds and fruits and the quantity of their production is reduced.
- In acute deficiency cases, shoots and plants may die, plants may become stunted and there may be little or no flowering.
- In older leaves, browning of tips (tip burns), marginal scorching (leaf scorch) or development of brown spots near the margins occurs. Scorched margins often turn sharply forward.
- Reduced production of grains occur.
- The lamina of broad leaved plants curl backward towards the under surface or roll forwards towards the upper surface parallel to the midrib.

Calcium (Ca)

Calcium is a component of plant cell walls and regulates cell wall formation. It occurs in soil with a variety of minerals. Ca is not mobile and is not translocated in the plant, so symptoms first appear on the younger leaves, leaf tips and near the growing points of stem and roots. In some crops, younger leaves may be cupped and crinkled, with the terminal bud deteriorating. Buds and blossoms fall prematurely in some crops.

Its deficiency is not often observed in plants. However, insufficient Ca can cause young leaves to become distorted and turn abnormally dark green. Leaf tips often become dry or brittle and will eventually wither and die. Stems are weak and seed germination is poor. Without adequate Ca, newly emerging leaves may stick together at the margins, which causes tearing as the leaves expand and unfurl. This may also cause the stem structure to be weakened.

Its deficiency symptoms studied in cereals and quinoa plants are as under:

- The growing tips of roots and leaves turn brown and die.
- Young leaves may be severely distorted with the tips hooked back and the margins curled backwards or forward or rolled.
- Roots poorly developed, lack fibre and may appear gelatinous.

Magnesium (Mg)

Magnesium is the central molecule in chlorophyll and is an important co-factor for the production of ATP. Mg is a mobile element and part of the chlorophyll molecule, the deficiency symptom of interveinal chlorosis first appears in older leaves. Leaf tissue between the veins may be yellowish, bronze or reddish, while the leaf veins remain green. Mg deficiency symptoms occur most frequently in acid soils and soils receiving high amounts of calcium or K fertilizers. Its deficiency symptoms include interveinal chlorosis and yellow or reddish-purple leaf margins, while the midrib remains green.

Its deficiency symptoms noticed in quinoa plants and other cereals are as under:

- In severe cases, symptoms may appear on the older leaves and proceed systematically towards the younger leaves and cause premature leaf drop.

- Severely affected leaves may wither and shed or abscise without the withering stage. Defoliation may be quite severe.
- Leaves sometimes develop necrotic spots.
- Stem becomes yellowish green, often hard and woody.

Sulfur (S)

As sulfur is an essential constituent of certain amino acids and proteins, its deficiency results in the inhibition of protein and chlorophyll synthesis. Sulfur deficiency symptoms can be difficult to diagnose as effects can resemble symptoms of N and Mo deficiencies. In contrast to N or Mo deficiency, however, S deficiency symptoms initially occur in younger leaves, causing them to turn light green to yellow. In later growth, the entire plant may be pale green. In some plants like in citrus, the older leaves may show symptoms first. However, deficiency is not commonly found in most plants. There are no characteristic spots or stripes. Additionally, plants deficient in Sulphur tends to be spindly and small, and stems are often thin.

Sulphur deficiency symptoms noticed in quinoa plants and other cereals are as under:

- Younger leaves are chlorotic with evenly, lightly coloured veins.
- Growth rate is retarded and maturity is delayed.
- Plant stems are stiff, thin and woody.
- Symptoms may be similar to N deficiency and are most often found in sandy soils that are low in organic matter and receive moderate to heavy rainfall.

Molybdenum (Mo)

Molybdenum is needed for enzyme activity in the plant and for nitrogen fixation in legumes. Due to this inter-relationship and role of Mo in assimilation of N in the plants, Mo deficiency symptoms often resemble N deficiency symptoms with stunted growth and chlorosis occurring in legumes. In tomato, chlorosis appears in the form of mottling in the older leaves. Other symptoms of Mo deficiency include pale leaves that may be scorched, cupped or rolled. Leaves may also appear thick or brittle and will eventually wither, leaving only the midrib. Mo deficiency can be common in nitrogen-fixing legumes.

Its deficiency symptoms noticed in quinoa plants and other cereal crops are as under:

- Older and middle leaves become chlorotic and the leaf margins roll inwards.
- In contrast to N deficiency, necrotic spots appear at the leaf margins because of nitrate accumulation.
- Defficient plants are stunted and flower formation may be restricted.
- Severe deficiency cause failure of grain formation.

Boron (Bo)

Plants suffering from Boron deficiency exhibit chlorotic young leaves and death of the main growing point (terminal bud). In addition to chlorosis, leaves may develop dark brown, irregular lesions that will progress to leaf necrosis in severe cases. Whitish-yellow spots may also form at the base of leaves. Due to disturbances in cell wall growth, leaves and stems of boron deficient plants become brittle and distorted, and leaf tips tend to thicken and curl. Affected plants grow slowly and appear stunted as a result of shortened internodes (stem segment between points where leaves are attached). Because Boron tends to accumulate in reproductive tissues, flower buds may fail to form or are misshapen and pollination and seed viability is usually poor in Boron deficient plants.

Deficiency symptoms of boron noticed in quinoa plants and cereals are as under:

- Boron deficiency causes stunted growth of the plants.
- The leaves tend to be thickened and may curl and become brittle.
- Yellowing of upper leaves and poor flowering.
- Development of hollow pith in stems.
- Boron deficiency exhibit chlorotic young leaves and death of the main growing point (terminal bud).

Iron (Fe)

Iron deficiency reduces chlorophyll production and is characterized by interveinal chlorosis with a sharp distinction between veins and chlorotic areas in young leaves. As the deficiency develops, the entire leaf becomes whitish-yellow and progresses to necrosis. Plant growth is slow. When viewed from a distance, Fe deficient field exhibit irregularly shaped yellow areas, especially where the sub-soil is exposed at the surface. Iron deficiency usually observed in alkaline or over-limed soils.

Deficiency symptoms of iron noticed in quinoa plants and other cereals are described hereunder:

- Interveinal white chlorosis of the leaves occurs and younger leaves are always most severely affected. Principal veins may remain green. The youngest leaves may be white, because iron, like Mg, is involved in chlorophyll production.
- Chlorosis may produce a mottled pattern or the leaf may show complete bleaching or often become necrotic.
- In extreme conditions, scorching of leaf margins and tips may occur.

Zinc (Zn)

Zn deficient leaves generally display interveinal chlorosis, especially midway between the margin and midrib, producing a striping effect; some mottling may also occur. Chlorotic areas can be pale green, yellow or even white. Interveinal chlorosis occurs on younger leaves, similar to Fe deficiency. However, Zn deficiency is more defined, appearing as banding at the basal part of the leaf, where as Fe deficiency results in interveinal chlorosis along the entire length of the Leaf. Severe Zn deficiency causes leaves to turn gray-white and fall pre-maturely or die. Zn deficient plants generally exhibit severe stunting. Flowering and seed set is also poor in affected plants.

Zink deficiency symptoms noticed in quinoa plants and cereals have been described as under:

- Leaves show chlorosis or necrosis and become leathery.
- Plants show resetting and premature shedding.
- Whitish chlorotic streaks between the veins are found in older leaves.
- Leaf margins are distorted, become twisted or wavy which later curl and look sickle shape.

Copper (Cu)

Copper deficient plants display chlorosis in younger leaves, stunted growth, lodging and in some cases, melanosis (brown discolouration). Copper deficient plants are prone to increased disease, specifically ergot (a fungus causing reduced yield and grain quality). In the field, Cu deficiency symptoms occur in irregular patches with melanosis being the most obvious symptom, particularly in wheat stands. Similar to Zn deficiency, forage that is deficient in Cu can cause reduced reproductive efficiency in cattle. In trees, multiple sprouts occur at growing points, resulting in a bushy appearance. Young leaves becomes bleached and eventually there is defoliation and dieback of twigs. In forage grasses, young leaf tips and growing points are affected first. The plant is stunted and chlorotic.

Its deficiency symptoms studied in cereals and quinoa are as under:

- Grain production and fill is often poor and under severe deficiency, grain heads may dwarf, distorted not even form.
- Eexcessively late tillering in grain crops and delayed maturity.
- Both vegetative and reproductive growth are reduced with distortion of the younger leaves and possible necrosis of the apical meristem.
- Wilting of terminal shoots occur which is followed by frequent death.
- Leaf colour is often faded due to reduction in carotene and other pigments.

Manganese (Mn)

Chloroplasts are the most sensitive of cell organs to Mn deficiency. As a result, a common symptom of Mn deficiency is interveinal chlorosis in young leaves. However, unlike with Fe deficiency, there is no sharp distinction between veins and interveinal areas, but rather a more diffuse chlorotic effect. In legumes, necrotic areas develop on the cotyledons, a symptom known as marsh spots. Two well-known Mn deficiencies in arable crops are grey speck in oats and marsh spot in peas. White streak in wheat and interveinal brown spot in barley are also symptoms of Mn deficiency.

Its deficiency symptoms studied in cereals and quinoa are as under:

- Symptoms first appear as chlorosis in young tissues. Unlike Fe chlorosis symptoms in dicots, Mn chlorosis shows up as tiny yellow spots.
- Greenish-grey specks appear at the lower base of younger leaves. The specks may eventually become yellowish to yellow-orange.
- Severe deficiency cause white streak.

Chloride (Cl)

Until recently, little information was documented on Cl deficiencies, as symptoms were often misdiagnosed as physiological leaf spot with visual symptoms observed in winter wheat and durum wheat cultivars. Plants with insufficient Cl show chlorotic and necrotic spotting along leaves with abrupt boundaries between dead and live tissue.

Wilting of leaves at margins and highly branched root systems are also typical Cl deficiency symptoms, found mainly in cereal crops. Cl deficiencies are highly cultivar specific and can be easily mistaken for leaf diseases. Deficiency seldom occurs because Cl is found in the atmosphere and rainwater. In quinoa, *Chlorosis* of younger leaves and wilting of the plant occurs.

11.7 TIME AND METHOD OF FERTILIZER APPLICATION

To obtain the maximum benefit from the fertilizers during quinoa cultivation, it is most essential that fertilizers should be applied at the proper time and at the proper place. The fertilizers to be applied possess different qualities with regard to solubility in water and movement into the soil solution. The soils are of different nature i.e., sandy to clayey which governs the movement of applied fertilizers. Again, the requirement of plants for different plant nutrients varies in relation to their stage of growth. For example, nitrogen is absorbed by the plants throughout the growth period, while phosphorus is absorbed at the faster rate during the early growth period. Thus, the time and method of fertilizer application will vary in relation to the nature of fertilizer, the soil type and the difference in nutrient requirement and nature of field crops.

Besides these three main factors, the following general principles governing the selection of proper time for application of fertilizers are listed below.

Principles governing fertilizers application:

(i) Nitrogen is required throughout growth of the plants. As such, it is absorbed by the plant at the same rate as that of its growth. All plants grow at a slow rate in the beginning. Then follows a rapid increase in the growth rate. Near the harvest time, the rate of gwowth is again slows down. Accordingly, nitrogen is taken up by the plant slowly in the beginning, rapidly during the grand growth period and again slowly as it nears maturity. In other words, the nitrogen requirement of growing plant is less in early stages of growth, maximum during its grand growth period and very slow at the subsequent stages up to harvest. It is, thus, seen that nitrogen is required throughout the growth period.

(ii) Nitrogenous fertilizers are soluble in water. They are mobile and move rapidly in all directions within the soil. As such nitrogen is easily lost through leaching. Since the nitrogen is required throughout growth period and nitrogenous fertilizers are lost through leaching, it is better not to apply too much nitrogenous fertilizers at one time, but to apply in split doses throughout the gwowth period. This will supply nitrogen to the growing plants during the entire growth period and the plants will not suffer from nutritional defficiency.

(iii) Phosphorus is required during the early root development and early plant growth. As such crop plants utilize 2/3 of the total requirement of phosphorus when the plants accumulate 1/3 of their dry weight. In other words, plants require more of phosphorus during the early growth period.

(iv) All phosphatic fertilizers release phosphorus slowly for plant growth. This is true even of superphosphate which contains mono-calcium phosphate or water soluble P_2O_5. On application of super phosphate to the soil, water soluble P_2O_5 becomes immediately slightly insoluble or is converted into dicalcium phosphate or citrate soluble P_2O_5. In this form, phosphorus becomes available to plants slowly.

On the one hand, phosphorus is required in greater quantities during the early growth period, while on the other, all phosphatic fertilizers become available to the growing plants slowly. As such, in quinoa, it is recommended that the entire quantity of phosphatic fertilizers should be applied before sowing / planting.

(v) Potash behaves partly like nitrogen and partly like phosphorus. From point of view of the rate of absorption, it is like nitrogen, being absorbed upto the harvesting stage. But potassic fertilizers, like phosphatic fertilizers, become available slowly. As such, it is always advisable to apply the entire quantity of potash at sowing time in quinoa cultivation.

(vi) Leaching is greater from sandy soils than from heavier textured soils, like clay soils. This means that more frequent application or split application of nitrogenous fertilizers and sometimes potassic fertilizers, is desirable on sandy soils.

11.8 RECOMMENDED DOSE OF FERTILIZERS FOR QUINOA

On the basis of guiding principles and field experiments, fertilizer application in quinoa should be as follows:

- Water soluble nitrogenous fertilizers should be applied in two split doses to quinoa crop which have 4 to 5 months duration. Nitrogenous fertilizers are applied in 3 splits to crops of 9 to 12 month duration and in 4 to 5 splits when crops are of still longer duration.
- On sandy / lighter soils, more frequent of split application of nitrogenous fertilizers is desirable, compared to heavy textured soils like clayey soils. This is important for reducing nitrogen losses due to leaching.
- The entire quantity of water soluble phosphatic fertilizers should be applied in one dose at sowing time. In acid soils, it is advisable to apply bone meal or rock-phosphate a week of fortnight prior to sowing.
- Potassic fertilizers should be applied in one dose at planting time.
- Nitrogen promotes strong leaf and stem growth and dark green colour. Nitrogen responds strongly to the application of nitrogen fertilizer. Grain yield show increasing trend with increasing N rates. The highest yield of quinoa was repcorded at 80 Kg nitrogen per hectare. At higher dose of nitrogen beyond this level, the quinoa yield was declined due to slow maturity and intense lodging. Nitrogen application is known to increase seed yield as well as the

protein content of the seeds. When quinoa is grown in heavily fertilized fields, it can accumulate nitrates in high concentrations.. Nitrogen applications have also significant impact on protein content of quinoa seed. The experiments conducted by author under Integrated Quinoa Development Project revealed that every kg of ammonium nitrate applied per ha, would increase the seed protein content by 0.1 percent.

- Phosphorus (P) promotes root and early plant growth, including setting blossoms and developing fruit and seed formation. Potassium (K) promotes plant root vigor and disease and stress resistance and enhances flowering. Heavy doses of phosphorus and potash are known to increase vegetative growth without any increase in seed yield. The maximum grain yield was recorded with the application of 40 Kg phosphorus and 45 Kg potash per hectare. Application of excess chemical/synthetic fertilizers should be avoided in quinoa cultivation with a view to maintain quality of grains. It can damage roots and/or reduce the availability of other elements.

11.9 DIAGNOSING NUTRIENT TOXICITIES IN QUINOA PLANTS

Over dose of nutrients cause visual symptoms. Macro-nutrient (N, P and K) toxicities most often occur as a result of the over-application of fertilizers or manure. Secondary macro-nutrient (Ca, Mg and S) toxicities are rare and toxic effects on quinoa plants have not been documented. Micro-nutrient toxicities can occur and are likely caused by over-application of fertilizer or manure, using irrigation water high in micro-nutrients or salts or in areas where soil micro-nutrient concentrations are abnormally high (i.e., areas exposed to mining activity or high metal minerals in sub-soil). In addition, high amounts of non-essential elements, such as arsenic (As), cadmium (Cd) and lead (Pb) can be directly toxic to plants and livestock or cause a nutrient imbalance in the plant, in which essential nutrient deficiencies or toxicities may possibly occur.

N, P and K

Plants with excess N turn a deep green colour and have delayed maturity. Due to positive effect of N on vegetative growth, excess N results in tall plants with weak stems, possibly causing lodging. New growth will be succulent and plant transpiration high (low water use efficiency). N toxicity is most evident under dry conditions and may cause a burning effect. Plants fertilized with ammonium (NH_4^+) based fertilizers may exhibit NH_4^+ toxicity, indicated by reduced plant growth, lesions occurring on stems and roots and leaf margins rolling downward, especially under dry conditions.

Excess P indirectly affects plant growth by reducing Fe, Mn and Zn uptake; thus, potentially causing deficiency symptoms of these nutrients to occur. Zn deficiency is most common under excess P conditions. Due to a cation imbalance, K toxicity can cause reduced uptake.

Micro-nutrients:

For many crops, the range between deficiency and toxicity is narrower for micro-nutrients than macro-nutrients. This is particularly true for Boron in which the average sufficiency and toxicity ranges for various crops overlap one another i.e., 10–200 ppm is sufficiency range and 50–200 ppm is toxicity range. Boron toxicity results in chlorosis followed by necrosis. Symptoms begin at the leaf tip and margins, and spread toward the midrib. As the toxicity progresses, older leaves will appear scorched and fall prematurely.

Other micro-nutrients causing potential toxicity symptoms include Cu, Mn, Mo, Ni and Zn. Studies suggest excess Cu will displace Fe and other metals from important areas in the plant, causing chlorosis and other Fe deficiency symptoms, such as stunted growth, to appear. High Ni concentrations can also cause Fe to be displaced. Interveinal chlorosis may appear in new leaves of Ni toxic plants and growth may be stunted. Mn toxicity symptoms are generally characterized by blackish-brown or red spots on older leaves and an uneven distribution of chlorophyll, causing chlorosis and necrotic lesions on leaves. While Mo toxicity does not pose serious crop problems (crops may appear stunted with yellow-brown leaf dis-colourations), excess amounts of Mo in forage have been found to be toxic to livestock. Zn toxicity is not common, but can occur on very saline soils. Symptoms include leaves turning dark green, chlorosis, interveinal chlorosis and a reduction in root growth and leaf expansion. Excess Zn may induce Fe deficiency.

Nutrient deficiency and their over dose may cause decrease in crop health and productivity and may result in the appearance of unusual visual symptoms. General deficiency symptoms include stunted growth, chlorosis, interveinal chlorosis, purple or red discolouration and necrosis. Deficiencies of mobile nutrients first appear in older, lower leaves, whereas deficiencies of immobile nutrients will occur in younger, upper leaves. Nutrient toxicity is most often the result of over-application, with symptoms including abnormal growth (excessive or stunted), chlorosis, leaf discolouration and necrotic spotting. When in excess, many nutrients will inhibit the uptake of other nutrients, thus potentially causing deficiency symptoms to occur as well. Therefore, for good crop, good yield and excellent quality of quinoa, fertilizers should be applied at proper time, in proper dose and by proper method to avoid nutrient deficiencies and toxicities.

Chapter

12

HEAVY METALS – CONTAMINATION AND QUALITY DETERIORATION OF QUINOA PRODUCE

12.1 INTRODUCTION

Heavy metals are natural components of the Earth's crust and are usually present in all environmental matrices. However, the concentration of different heavy metals has increased several fold in some ecosystems as a result of anthropogenic activities. Contamination of environmental Heavy metals has continued to gain global attention, due to toxicological risks to human health posed by such metals present in different foods. Although metallic elements are often essential for living organisms, but, they become toxic when present at high concentrations. The rapid increase in human population, coupled with haphazard industrialization, technological advancement and application of fertilizers containing heavy metals creating many serious health problems around the world. In the past few decades, the concentration of heavy metals in soil and surface waters has increased and now constitutes a potential threat to terrestrial and aquatic biota and to humans by entering in the food chain. Due to the widespread presence of heavy metals in the environment, their residues also reach and are assimilated into food crops.

The demand of quinoa is increasing day by day due to its high value protein, excellent nutritional status and high medicinal values. Therefore, assurance of quality quinoa production is essential for its safe and healthy uses. The quality of quinoa may detereriorate when heavy metal containing fertilizers are used during its cultivation and they affect both plant growth attributes and production of secondary metabolites. High levels of heavy metal contamination of quinoa during its cultivation may suppress or stimulate synthesis of secondary metabolites in quinoa plants. Moreover, heavy metals may play an important role in triggering plant genes to alter the nature of secondary plant metabolites, although the exact mechanism by which this happens remains unclear. Oxidative stress induced by heavy metals triggers signaling pathways that affect production of specific plant metabolites. In particular, reactive oxygen species (ROS), generated during heavy metal stress, may cause lipid peroxidation that stimulates formation of highly active signaling compounds capable of triggering production of bioactive compounds (secondary metabolites). As usual, research is needed to find out and clarify the mechanism by which heavy metals affect secondary metabolite production in quinoa and other plants.

Plants do not deal with metals in their elemental forms; they are not accessible to plants. Metals are only available to them in solution and it is necessary for metals to react with other elements and form compounds before they can be solubilised.

Once such a chemical compound is formed, the density of the metal does not play any role. The correlation between the density of a metal and its physiological or toxicological effects on chemical properties of different compounds of quinoa is not clear so far.

12.2 METALS

Metals may be defined by the physical properties of the elemental state as elements with metallic luster, the capacity to lose electrons to form positive ions and the ability to conduct heat and electricity, but they are better identified by consideration of their chemical properties. The term is used indiscriminately by non-chemists to refer to both the element and compounds.

Metals are often characterised and distinguished from non-metals by their physical properties – the ability to conduct heat and an electrical resistance that is directly proportional to temperature, malleability, ductility and even lustre. These properties, especially that of a temperature dependent conductivity, at least allow us to define what a metal is in contrast to Heavy metals such as cadmium, lead and mercury are common air pollutants and are emitted (predominantly into the air) as a result of various industrial activities.

The Metals fall naturally into following two groups:

Light metal

A very imprecise term used loosely to refer to both the element and its compounds. It has rarely been defined, but the term is mostly applied to metals of density less than 4 $/cm^{-3}$.

Heavy metals

"Heavy metal" is a collective term for metals of high atomic mass, particularly those transition metals that are toxic and cannot be processed by living organisms. Metals with densities above $7/cm^{-3}$ are called generally heavy metals. This definition is meant to suggest that the density of a heavy metal is high, but their physical property is quite meaningless in the context of plants and other living organisms.

12.3 FUNCTIONAL CLASSIFICATION OF METALS

- **Essential metal:** Broadly, one which is required for the complete life cycle of an organism, whose absence produces specific deficiency symptoms relieved only by that metal and whose effect should be referred to a dose–response curve. The term is often used mis-leadingly since it should be accompanied by a statement of which organisms show a requirement for the element. Again, it is used loosely to refer to both the element and its compounds.
- **Beneficial metal:** An old term, now largely dis-used, which implied that a non-essential metal could improve health. It is another term that has been used loosely to refer to both the element and its compounds.

- **Toxic metal:** This term has been used loosely to refer to both the element and its compounds. The fundamental rule of toxicology is that all substances, including carbon and all other elements and their derivatives, are toxic given a high enough dose. The degree of toxicity of metals varies greatly from metal to metal and from organism to organism. Pure metals are rarely, if ever, very toxic, except as very fine powders, which may be harmful to the lungs from whatever substance they may originate. Toxicity should be defined by reference to a dose–response curve for the species under consideration.

1.4 IMPORTANT HEAVY METALS

- **Cobalt:** Cobalt is a naturally occurring element that exists in several forms. All cobalt used in industries is imported or obtained by recycling scrap metal that contains cobalt. Cobalt is used in alloys (mixtures of metals) and as a drier for paint and porcelain enamel used on steel bathroom fixtures, large appliances, and kitchenware. Radio-active cobalt is used to sterilize medical equipment and food and to treat cancer and anemia.
- **Chromium:** Chromium is a element of naturally occurring in rocks, animals, plants, soil and volcanic dust and gases. Chromium is present in the environment in several different forms. Some forms of chromium occur naturally in the environment and are essential nutrients in the human diet. Industrial processes generally produce other forms of chromium.
- **Cesium:** Cesium is a naturally occurring element in rocks, soil and dust at low concentrations. Cesium metal and its compounds have few commercial uses. Sometimes, it is used to absorb for residual gas impurities in vacuum tubes and as a coating in tungsten filaments or cathodes of the tubes. Naturally occurring cesium occurs in the environment mostly from the erosion and weathering of rocks and minerals. The mining and milling of certain ores can also release cesium to the air, water and soil. The level of cesium in air and water is usually low.
- **Cadmium:** Cadmium is a metallic element that occurs naturally in the earth's crust. All soils and rocks contain some cadmium. Cadmium has many uses, including batteries, pigments, metal coatings and plastics. Kidney and bone are the critical target organs with regard to environmental exposure. Cadmium is accumulating in soils and catchments under certain environmental conditions, thus increasing the risk of future exposure through food. Therefore, in view of the narrow margin of safety, every effort should be made to make further reductions regarding cadmium emission into the atmosphere and other types of cadmium input into soil. Food is the main source of cadmium exposure in the general population, representing 90% of the total intake in non-smokers. In heavily contaminated areas, dust resuspension can constitute a substantial part of the crop contamination and exposures via inhalation and digestion. Studies on the cadmium balance in the top layers of arable soils indicate that the input of this heavy metal still exceeds its removal.
- **Lead:** Lead is a naturally occurring metal found in small amounts in the earth's crust. Lead can be found in all parts of our environment. Much of it comes

from human activities that include mining, manufacturing and burning fossil fuels. Lead has many different uses. It is used to make batteries, ammunition, metal products (solder and pipes) and devices to shield X-rays. Because of health concerns, the amount of lead used in gasoline, paints, ceramics and pipe solder has been dramatically reduced in recent years.

The relative contributions of sources of pollution differ depending on local conditions. Food is the predominant source of lead uptake in the general population. Ingestion of contaminated soil, dust and old lead-based paint due to hand-to-mouth activities may also be important regarding lead intake in infants and young children. When tap-water systems with leaded pipes are used, lead intake via drinking-water can be an important source, especially in children. Inhalation exposure may be significant when lead levels in the air are high. Although uptake via plant roots is relatively small, rising lead levels in soils over the long term are a matter of concern and should be avoided because of the possible health risks of low-level exposure. Lead emissions to the atmosphere should, therefore, be kept as low as possible.

- **Barium:** Barium is a silvery-white metal found in nature. Barium compounds are used to make paint, bricks, tiles, glass and rubber; used by the oil and gas industries in drilling muds; and sometimes used by doctors to perform medical tests.
- **Arsenic:** Arsenic is a naturally occurring element widely distributed in the earth's crust. It combines with oxygen and other elements to form inorganic arsenic compounds. Inorganic arsenic compounds have various industrial purposes, including the smelting of other metals, such as lead, manufacture of some types of glass and formulation of pesticides and fungicides. Inorganic arsenic compounds also can be used in pressure-treated wood.
- **Mercury:** Mercury is a naturally occurring metal that has several forms. One form is metallic mercury, a shiny, silver-white, odorless liquid. Metallic mercury is used in thermometers, dental fillings and batteries. Mercury combines with other elements to form inorganic mercury compounds or "salts," which are usually white powders or crystals. Mercury salts are sometimes used in antiseptic creams and ointments. Methyl mercury is a potent neurotoxic chemical.

Emissions of mercury to the air from both anthropogenic and natural sources are in inorganic forms that can be converted biologically to methyl mercury in soil and water. Methyl mercury bio-accumulates and enters in the human body readily via the dietary route. Concentrations of inorganic mercury species in surface water and ground water are generally well below those known to cause adverse health effects from drinking water.

Un-born children (i.e., fetuses) are the most susceptible. The exposure being mainly from fish in the diet of the mother. Populations consuming large amounts of fish, or consuming contaminated fish, the intake of methyl mercury may reach hazardous levels. Thus, reducing the concentrations in fish should be given a high priority. Methyl mercury is also excreted in mothers' milk.

- **Selenium:** Selenium is a metal commonly found in rocks and soil and in low levels in water. It is an essential nutrient in the human diet. The major man-made source of selenium is burning coal. Selenium also is used in paints, fertilizers and dandruff shampoos.
- **Nickel (Ni):** Nickel is a very abundant element. Pure nickel is a hard, silvery-white metal. In the environment, it is found in all soils and is emitted from volcanoes. The industries with particular exposure to nickel are coin and jewelary making; fossil fuel combustion; welding; making rechargeable (nickel-cadmium) batteries; and production of glass bottles. Small nickel particles in the air settle on the ground or are taken out of the air in rain. Much of the nickel in the environment is found in soil and sediments. Nickel does not appear to collect in fish or animals used for food, but plants that are grown in nickel-rich soil can collect it.

It is required by plants for proper seed germination and is beneficial for metabolism of nitrogen in legumes and other plants in which ureides (compounds derived from urea) are important in metabolism. It is the metal component in urease, an enzyme that catalyzes the conversion of urea to ammonium. Ni deficiency symptoms include chlorosis and interveinal chlorosis in young leaves that progress to plant tissue necrosis. Other symptoms include poor seed germination and decreased crop yield.

- **Manganese**: Manganese is a naturally occurring metal in many type of rocks. It exists naturally in rivers, lakes and underground water. Manganese is a trace element essential for good health. It can be found in several food items, including grains and cereals and is found in high amounts in other foods such as tea.
- **Molybdenum:** Molybdenum is a naturally occurring metal that is an essential dietary nutrient. It is used in the petroleum industry and by the steel industry to make steel alloys. Molybdenum increases the strength and heat resistance of these alloys.
- **Uranium:** Uranium is a naturally occurring radioactive element. It is a normal part of rocks, soil, air and water. Natural uranium is a mixture of three types or isotopes, called U-234 (^{234}U), U-235 (^{235}U) and U-238 (^{238}U). All three types of uranium are the same chemical, but they have different radioactive properties. U-238 is by far the most common type of uranium on earth and is the least radioactive. U-234 is the most radioactive type of uranium. U-235 is used as a fuel in power plants and weapons. The amount of U-234 and U-235 in natural uranium is increased through a process called enrichment.
- **Thallium:** Thallium is a bluish-white metal found in trace amounts in the earth's crust. In the past, it was obtained as a by-product of smelting other metals. Thallium is used mostly in manufacturing electronic devices, switches and closures, mainly for the semi-conductor industry. It is used in making special glass and for certain medical procedures. In the past, thallium was used as a rodenticide.

- **Antimony:** Antimony is a silvery-white metal found in the earth's crust. Most antimony is brought into the United States from other countries for processing. Antimony is used in lead storage batteries, sheet and pipe metal, paints, ceramics and fireworks. It is also added to textiles and plastics to prevent them from catching fire.

12.5 TOXIC EFFECTS OF HEAVY METALS

The term "heavy metal" is linked in many people's minds to metals (or their compounds) that are toxic. However, this is a feeling rather than a conclusion based on scientific evidence. Following two facts should be kept in mind:

1. The effect of any substance on a living system is always dependent on the concentration of it available to cells. Thus, there are no substances that are always toxic. What we need to evaluate toxicity are dose–response data; i.e., quantitative dose–response relationships.
2. Several metal ions are crucial to the metabolism of cells at low concentrations but are toxic at high concentrations, resulting in bell-shaped dose–response relationships. These metals are sometimes called micronutrients.

- **Toxicity of heavy metals in biological systems:** Heavy metal is not toxic, it is only toxic when its concentration in the plant exceeds a certain threshold (it is the dose that makes the effect). This is especially important to the second fact: that some elements, called micro-nutrients, have essential functions in plant cells. This has been shown for Co, Cu, Fe, Mn, Mo, and Zn. Only when the internal concentration exceeds a certain threshold, they demonstrate toxic effects and then they are commonly termed "heavy metals".

Micronutrients are essential for biosynthesis, growth, nucleic acids, growth substances, chlorophyll and secondary metabolites, carbohydrates and lipids as well as for stress resistance. A supply of micronutrients is also essential for the integrity of membranes. Some of the non-essential heavy metals have a stimulating or inducing effect when they are applied at very low concentrations (these are termed low concentration stressors).

(i) **Effect on photosynthesis:** As already mentioned above, inhibition of photosynthesis is one effect that most of the heavy metals have in common when present at toxic concentrations. It is a very sensitive response. Measuring the photosynthetic activity is a good screening method for detecting possible stress situations, perhaps including those involving heavy metals. Direct effects of heavy metals on light and dark reactions and indirect effects caused by them decreasing the photosynthetic pigment content are involved, as well as changes in stomata function. It seems that nearly all of the components of the photosynthetic apparatus are influenced by almost all heavy metals, including chlorophyll and carotenoid content, chloroplast membrane structure, light-harvesting and oxygen-evolving complexes, photosystems and constituents of the photosynthetic electron transport chain. Several enzymes involved in the Calvin cycle are also inhibited.

(ii) **Effect of heavy metals on mitochondrial respiration:** Some scientists consider that respiration increases under stress. However, this is more than an over simplification. It is true that in some plant species, the presence of some heavy metals at lower concentrations increases respiration. At toxic concentrations of heavy metals, respiration is usually inhibited.

- **Health effects of Heavy metals**

(i) Arsenic: People exposed to high levels of arsenic can have nausea and vomiting, diarrhea, anemia and low blood pressure. These symptoms may be followed by a feeling of "pins and needles" in the hands and feet (neuropathy). Chronic (long-term) exposure to arsenic can cause stomach ailments, headaches, fatigue, neuropathy, dark splotches on the skin and small "corns" or "warts" on the palms of the hands, soles of the feet and torso. People exposed to inorganic arsenic can have more cancer of the lung, skin, bladder, liver, kidney and prostate. Studies have not linked arsenic exposure to leukemia in adults or children.

(ii) Barium: The health problems from different barium compounds depend on how well the compound dissolves in water. Barium compounds that do not dissolve well in water, such as those used in medical tests, generally are not harmful. The barium compounds that do dissolve well in water can harm people. Eating food and drinking water with high levels of barium compounds that easily dissolve in water for a short time can cause stomach irritation, higher blood pressure, muscle weakness or damage to the liver, kidney, heart and spleen. We do not know the effects in people who have ingested low levels of barium over a long time, but lab animals that ate barium over a long period had high blood pressure and changes in the heart. The few animal studies that have been done to determine whether barium causes cancer have produced unclear results.

(iii) Cadamium: Breathing high levels of cadmium can damage the lungs. Eating food or drinking water with very high levels of cadmium can irritate the stomach, leading to vomiting and diarrhea. Exposure to lower levels of cadmium over a long time builds up cadmium in the kidneys and possibly leads to kidney disease. Other long-term effects are lung damage and fragile bones. Some studies have shown that work-place exposure to cadmium can cause lung and prostate cancer, but other studies have not shown this. According to the Environmental Protection Agency, cadmium probably causes lung cancer and possibly causes prostate cancer.

The main critical effects include increased excretion of low-molecular weight proteins in the urine (as a result of proximal tubular cell damage) and an increased risk of osteoporosis An increased risk of lung cancer has also been reported following inhalation exposure in occupational settings. The margin of safety between the present daily intake of cadmium in the diet and the intake that can result in effects is very narrow and, for highly exposed sub-populations, even non-existent. Population groups at risk include the elderly, people with diabetes and smokers. Women may be at increased risk because they have lower iron stores than men and, consequently, absorb more cadmium at the same level of exposure.

(iv) **Chromium:** All forms of chromium can be toxic at high levels, but some forms are more toxic than others. Breathing high levels of chromium can cause irritation to the nose, such as runny nose and nose bleeds. Ingesting large amounts of chromium can cause stomach upsets and ulcers, seizures, and kidney and liver damage. Skin contact with certain chromium compounds can cause skin ulcers. Some people are extremely sensitive to chromium and can have allergic reactions consisting of severe redness and swelling of the skin. One form of chromium compound can increase the risk of lung cancer, the EPA classifies it in air as a human carcinogen.

(v) **Cobalt:** Exposure to high levels of cobalt can harm human health. Some workers who breathed high levels of cobalt in the air in combination with tungsten (hard metal disease) developed lung problems, including asthma, pneumonia and wheezing. Studies have not consistently shown that exposure to natural (non-radioactive) cobalt causes cancer in humans. One study in workers with hard metal disease showed more deaths from lung cancer than in workers not exposed to cobalt dust. Another study linked exposure of workers to dust containing a mixture of tungsten and cobalt cause lung cancer.

(vi) **Lead:** Lead can affect almost every organ and system in the body. The most sensitive system, particularly in children, is the central nervous system which includes the brain. Slow development can occur in young children who have lead levels as low as 10 ug/dL. At hazardous levels, lead can decrease reaction time; cause weakness in fingers, wrists or ankles and affect the memory. Lead also can cause anemia, stomach problems and high blood pressure. Lead can cause cancer in animals, but not enough studies in humans have been done to show whether lead causes cancer in humans.

Lead is a well-known neurotoxin. Impairment of neuro-development in children is the most critical effect. Exposure in utero, during breast feeding and in early childhood may all be responsible for the effects. Lead accumulates in the skeleton and its mobilization from bones during pregnancy and lactation causes exposure to fetuses and breast-fed infants.

(vii) **Mnganese:** Some people exposed to very high levels of manganese for long periods of time in their work have developed mental and emotional disturbances and slow and clumsy body movements. Exposure to high levels of the metal can also cause respiratory problems and sexual dysfunction. No information is available about whether manganese causes cancer in humans. Male rats in lab exposed to high levels of manganese in food had slightly more pancreatic tumors. Male and female mice had slightly more thyroid tumors.

(viii) **Mercury:** The nervous system is very sensitive to all forms of mercury. Methyl mercury and metallic mercury vapours are more harmful than other forms because more mercury in these forms reaches the brain. Exposure to high levels of metallic, inorganic or organic mercury can permanently damage the brain, the kidneys and the developing fetus. Effects on brain functioning may cause irritability, shyness, tremors, changes in vision or hearing and memory problems.

Short-term exposure to high levels of metallic mercury vapours can cause problems, such as lung damage, vomiting, diarrhea, skin rashes and eye irritation. Not enough research has been done for us to know whether mercury causes cancer in humans. However, some forms of mercury cause cancer in lab animals.

(ix) **Molybdenum:** Molybdenum can irritate the eyes, nose and throats of exposed workers. Some workers exposed to molybdenum have more headaches, backaches and aching joints than workers not exposed to molybdenum. Some workers exposed to molybdenum developed lung disease, but this could not be linked to molybdenum because these workers were exposed to other metals as well. Molybdenum has not been linked to cancer in lab animals and no information is available about whether it causes cancer in humans.

(x) **Nickel:** A small amount of nickel is probably essential for humans, although a lack of nickel has not been found to affect human health. The most common health problem from nickel is an allergic reaction at the site of contact. People can become sensitive to nickel when jewelry or other items containing it touch their skin. Once a person is sensitized to nickel, further contact with it will produce a reaction. Some people who are sensitive to nickel have asthma attacks or conjunctivitis after exposure.

Workers who breathed large amounts of nickel can have chronic bronchitis, sinusitis, nasal polyps, asthma and nasalgand lung cancers. Today, levels of nickel in work place air are much lower than they were in the past and few workers now show symptoms of nickel exposure. Workers who accidentally drank water containing very high levels of nickel (100,000 times more than in normal drinking water) had stomach aches, shortness of breath, headaches and blood and kidney problems.

(xi) **Selenium:** Selenium compounds in the diet can be harmful at daily levels 5–10 times higher than the daily requirement. Too much dietary selenium produces nausea, fatigue, hair loss, deformed nails and a feeling of "pins and needles" in the hands and feet (neuropathy). Over-exposure to selenium in foods is unlikely to occur in the United States. Selenium compounds that touch the skin can cause redness and pain. High levels of selenium in workplace air can irritate the nose and throat. Selenium sulfide, a compound very different from the selenium compounds found in nature and foods, caused liver tumors in rats and lung tumors in mice that were fed high amounts daily. However, its use in dandruff shampoos is considered safe.

(xii) **Uranium:** Large amount of uranium can react with the tissues in the body and damage the kidneys. This damage results from the chemical properties of uranium, not from its radioactive properties. Its radioactive elements of uranium, the chance of getting cancer is greater for people who are exposed to enriched uranium. No cancer of any type has ever been seen in humans as a result of exposure to natural (non-enriched) uranium.

(xiii) **Thallium:** Exposure to high levels of thallium can cause health problems. Workers who breathe thallium over a long period can have problems of nervous system, such as numbness of the fingers and toes. Ingestion of large amounts of thallium contaminated food or water over a short time causes vomiting, diarrhea, temporary hair loss and problems with the nervous system, lungs, heart, liver and kidneys. It is still not clear about the problems from ingesting low levels of thallium over a long period of time. No information is available for people or animals about whether thallium exposure causes cancer.

(xiv) **Antimony:** Breathing high levels of antimony for a long time can irritate eyes of people and can cause heart and lung problems, stomach pain, diarrhea, vomiting and stomach ulcers. Ingesting large doses of antimony can cause vomiting. It is not clear about the problems caused by eating or drinking it and also whether antimony cause cancer in people. Long-term studies in lab animals found liver damage and blood changes when animals ingested antimony. In some lab studies in which the animals breathed high levels of antimony, the animals got lung cancer.

Chapter

13

TECHNOLOGIES FOR ORGANIC QUINOA PRODUCTION

13.1 INTRODUCTION

At the time of independence, India was a hungry due to low productivity of foodgrains. With the growth in population and increasing pressure on land, the need for intensive agriculture was recognized and Green Revolution was brought in. It was based on monoculture of high yielding varieties supported by chemical fertilizers and pesticides. The long-term unjudicious and unbalance use of chemical fertilizers and pesticides has an adverse effect to the environment, soil health, ground water and crop produce, and continous application of these chemicals leads to soil acidification, decreased the soil pH and reduces the availability of vital nutrients for plant growth. Chemical fertilizers and pesticides also leads to depletion of soil micro-nutrients and soil organic carbon, which influence the growth and quality of crop plants and health of human and animals.

As a consequence, many environmental problems, such as development of resistance in insect-pests to pesticides, resurgence of target and non target pests, destruction of beneficial organisms like honey bees, pollinators, parasitoids, predators etc have emerged and quality of food, fodder and feed has gone down due to residues of chemical pesticides and fertilizers. In fact, Green Revolution has come up against ecological and natural resources. Its classical packages of practice which include heavy doze of fertilizers, pesticides and of course high need of irrigations were not sustainable. Now, it has become necessary to explore non-chemical trajectories for sustainable agricultural growth, high quality food production and prevention of degradation of natural resources. Organic farming is certainly an answer for health security through production of healthy foods and clean environment.

13.2 ORGANIC FARMING

Organic farming has emerged as an important priority area globally in view of the growing demand of safe and healthy foods. It also has long term sustainability and concerns on environmental pollution associated with indiscriminate use of agro-chemicals. Though the use of chemical inputs in agriculture is inevitable to meet the growing demand for food in world, there are opportunities in selected crops like quinoa, amaranthus, chia seeds etc. and niche areas where organic production can be encouraged to tape the domestic export market through quality crop produce.

Indiscriminate use of synthetic fertilizers has led to the pollution and contamination of the soil, has polluted water basins, destroyed micro-organisms and friendly insects, making the crop more prone to diseases and reduced soil fertility. Demand is much higher than the availability. It is estimated that by 2020, to achieve the targeted production of 321 million tones of food grain, the requirement of nutrient will be 28.8 million tones, while their availability will be only 21.6 million tones being a deficit of about 7.2 million tones, depleting feedstock/fossil fuels (energy crisis) and increasing cost of fertilizers. This is becoming unaffordable by small and marginal farmers, depleting soil fertility due to widening gap between nutrient removal and supplies, growing concern about environmental hazards, increasing threat to sustainable agriculture. Besides above facts, the long term use of bio-fertilizers is economical, eco-friendly, more efficient, productive and accessible to marginal and small farmers over chemical fertilizers.

The organic food products are now getting better prospects for market and trade. The people world over choose over organic food due to better taste, better keeping quality and negligible residues of pesticides. There is a boost in international and domestic market of organic quinoa and other agricultural foods in India. The organic quinoa and others agri-products fetch a premium price compared with conventional products or those grown with chemical fertilizers / pesticides. Thus, production of quinoa organically is an important step towards its development in India and abroad.

13.3 RECOMMENDED ORGANIC FERTILIZERS FOR QUINOA

In organic crop production, organic manures / organic fertilizers are generally used as potential alternative to chemical fertilizers, which are not in conflict with nature. Such manures include composts, Farm Yard Manure, green manures, biogas digested slurry, vermi-compost, vermi-wash etc. These organic manures help in improving soil fertility and provide all essential nutrients to the plants in balanced proportion. Some of these manures like vermi-compost contain sufficient count of useful soil microbes, which help in increasing the fertility of soil leading sustainable crop production.

The best organic manures recommended for organic quinoa production have been described here under:

A. **Bio-Fertilizers:** The bio-fertilizers play a major role in enhancement of soil fertility, yield attributing characters and thereby final yield. In addition, their application in soil improves soil-biota and minimizes the sole use of chemical fertilizers. Biofertilizers add nutrients through the natural processes of fixing atmospheric nitrogen, solubilizing phosphorus and stimulating plant growth through the synthesis of growth promoting substances. Despite of promising results, bio-fertilizers have not got widespread application in quinoa production mainly because of the variable response of plant species or genotypes to inoculation depending on the bacterial strain used. Differential rhizosphere effect of crops in harbouring a target strain or even the nodulation of the bacterial nitrogen fixing and phosphate solubilizing capacity by specific root exudates may account for the observed differences. On the other hand, good competitive ability and high saprophytic competence are the major factors determining the success of a bacterial strain as an inoculant.

   'Biofertilizer' when applied to seed, plant surfaces or soil, colonizes the rhizosphere or the interior of the plant and promotes growth by increasing the supply or availability of primary nutrients to the host plant. Biofertilizers are not come under fertilizer's category which directly increase soil fertility by adding nutrients.

Bio-fertilizers can be grouped in different ways based on their nature and function. The main suitable bio-fertilizers for organic cultivation of quinoa are given below:

- **Azotobacter:** ***Azotobacter*** is an important and well known free living nitrogen fixing, aerobic and heterotrophic biofertilizer. It belongs to the family – ***Azotobacteriaceae.*** It is used as a Bio-fertilizer for all non-leguminous plants including quinoa, other cereals, millets, vegetables etc. Azotobacter cells are abundant in the rhizosphere region. The lack of organic matter in the soil is a limiting factor for the proliferation of Azotobaceter. *Azotobacter*s are present in neutral or alkaline soils and *A. chroococcum* is the most commonly occurring species in arable soils. *A. vinelandii, A. beijerinckii, A. agilis* and *A. macrocytogenes* are other reported species. The number of *Azotobacter* rarely exceeds of 104 to 105 g^{-1} of soil due to lack of organic matter and presence of antagonistic micro-organisms in soil. The bacterium produces anti-fungal antibiotics which inhibits the growth of several pathogenic fungi in the root region thereby preventing seedling mortality to a certain extent. The population of *Azotobacter* is generally low in the rhizosphere of the crop plants and in un-cultivated soils. The occurrence of this organism has been reported from the rhizosphere of a number of crop plants such as cereals, quinoa, rice, maize, sugarcane, bajra, vegetables and plantation crops.
- **Azospirillum:** ***Azospirillum*** is one kind of bacteria which is heterotrophic and associative in nature and known to fix the considerable quantity of nitrogen in non-leguminous plants. It belongs to the family ***Spirilaceae.*** In addition to their nitrogen fixing ability of about 20–40 kg ha^{-1}, they also produce growth regulating substances. The bacteria of Genus - *Azospirillum* are N_2 fixing organisms isolated from the root and above ground parts of a variety of crop plants. They are Gram negative, Vibrio or Spirillum having abundant accumulation of polybetahydroxybutyrate (70%) in cytoplasm. Five species of Azospirillum have been described to date *A. brasilense, A.lipoferum, A.amazonense, A. halopraeferens and A. irakense* but, worldwide distribution and benefits of inoculation have been proved mainly with the *A. lipoferum and A. brasilense.* The *Azospirillum* form associative symbiosis with many plants particularly with those having the C4-dicarboxyliac path way of photosynthesis, because they grow and fix nitrogen on salts of organic acids, such as malic, aspartic acid. Thus, it is mainly recommended for cereals, quinoa, maize, sugarcane, sorghum, pearl millet, cotton etc. The Azotobacter colonizing the roots not only remains on the root surface but also a sizeable proportion of them penetrates into the root tissues and lives in harmony with the plants. They do not, however, produce any visible nodules or out growth on root tissue. *Azospirillum lipoferum* and *A. brasilense* are primary inhabitants of soil, the rhizosphere and intercellular spaces of root cortex of graminaceous plants.

They perform the associative symbiotic relation with the graminaceous plants. The organism proliferates under both anaerobic and aerobic conditions but it is preferentially micro-aerophilic in the presence or absence of combined nitrogen in the medium. Apart from nitrogen fixation, growth promoting substance production (IAA), disease resistance and drought tolerance are some of the additional benefits due to *Azospirillum* inoculation.

- **Phosphate solubilizing bacteria (PSB):** It is an established fact that the efficiency of phosphate fertilizers is very low (15–20%) due to its fixation in acidic and alkaline soils and unfortunately both soil types are pre-dominating in India accounting more than 34% acidity affected and more than seven million hectares of productive land salinity/alkaline affected. Therefore, the inoculations with PSB and other useful microbial inoculants in these soils become mandatory to restore and maintain the effective microbial populations for solubilization of chemically fixed phosphorus and availability of other macro and micro-nutrients to harvest good sustainable yield of various crops. Very often micro-organisms are not as efficient in natural surroundings as one would expect them to be and therefore artificially multiplied cultures of efficient selected micro-organisms play a vital role in accelerating the microbial processes in soil.

Several studies have examined the ability of different bacterial species to solubilize insoluble inorganic phosphate compounds, such as tricalcium phosphate, dicalcium Phosphate, and rock phosphate. Among the bacterial genera with this capacity are *Pseudomonas, Bacillus, Rhizobium, Burkholderia, Achromobacter, Agrobacterium, Microccocus, Aereobacter, Flavobacterium* and *Erwinia.* There are considerable populations of phosphate solubilizing bacteria in soil and in plant rhizospheres. These include both aerobic and anaerobic strains, with a prevalence of aerobic strains in submerged soils. A considerably higher concentration of phosphate solubilizing bacteria is commonly found in the rhizosphere in comparison with non-rhizosphere soil. The soil bacteria belonging to the genera *Pseudomonas* and *Bacillus* and *fungi* are more common.

Phosphate solubilizing bacteria (PSB) are a group of beneficial bacteria capable of hydrolysing organic and inorganic phosphorus from insoluble compounds. P-Solubilization ability of the microorganisms is considered to be one of the most important traits associated with plant phosphate nutrition. It is generally accepted that the mechanism of mineral phosphate solubilization by PSB strains is associated with the release of low molecular weight organic acids, through which their hydroxyl and carboxyl groups chelate the cations bound to phosphate, thereby converting it into soluble forms. In addition, some PSB produce phosphatase like phytase that hydrolyse organic forms of phosphate compounds efficiently. One or both types of PSB have been introduced to Agricultural community as phosphate biofertilizer. Phosphorus (P) is one of the major essential macronutrients for plants and is applied to soil in the form of phosphate fertilizers. However, a large portion of soluble inorganic phosphate which is applied to the soil as chemical fertilizer is immobilized rapidly and becomes unavailable to plants. Currently, the main purpose in managing soil phosphorus is to optimize crop production and minimize P loss from soils. PSB have attracted the attention of agriculturists as soil inoculums to improve the plant growth and yield. When PSB used with rock phosphate, it can save about 50 percent of the crop requirement of phosphatic fertilizer. The use of PSB as inoculants increases the P uptake by plants. Simple inoculation of seeds with PSB gives crop yield responses equivalent to 30 kg P_2O_5 ha^{-1} or 50 percent of the need for phosphatic fertilizers. Currently, different strains of these bacteria has been identified for using in biofertilizers, among them *Pantoea agglomerans (P5), Microbacterium laevaniformans (P7) and Pseudomonas putida (P13)* strains are highly efficient insoluble phosphate solubilizers.

Several soil bacteria and fungi, notably species of *Pseudomonas, Bacillus, Penicillium,* and *Aspergillus* etc. secrete organic acids and lower the pH in their vicinity to bring about dissolution of bound phosphates in soil. Currently, phosphate solubilizers are manufactured by State Agricultural Universities and some private enterprises and sold to farmers through governmental agencies. These appear to be no check on either the quality of the inoculants marketed in India or the establishment of the desired organisms in the rhizosphere.

Methods of application of bio-fertilizers

Bio-fertilizers are used generally for seed treatment or seed inoculation, root dipping of seedling and soil application as described hereunder:

- **Seed treatment:** One packet of the inoculant (200 g) is sufficient to treat seeds of one acre quinoa crop. One packet of the inoculant is mixed with 200 ml of slurry. The seeds required for an acre are mixed in the slurry so as to have a uniform coating of the inoculant over the seeds and then shade dried for 30 minutes. The shade dried seeds should be sown within 24 hours.
- **Seedling root dip:** This method is used where seedlings of quinoa are planted. Two packets of the inoculant are mixed in 40 litres of water. The root portion of the seedlings required for an acre is dipped in the solution / mixture for 5 to 10 minutes and then transplanted.
- **Application in field:** Four packets of the bio-fertilizer / inoculant is mixed with 20 kg of dried and seived / powdered Farm Yard Manure. This mixture is then broadcasted in one acre field just before transplanting of quinoa seedlings.

Precautions in use of bio-fertilizers

- Bacterial inoculants should not be mixed with insecticide, fungicide, herbicide and chemical fertilizers.
- Seed treatment with bacterial inoculant is to be done at last when seeds are treated with fungicides.

B. Vermi-compost

Charles Darwin (1837 1881), a father of **"Earthworm Bio-technology"** first showed the significance of earth-worms in agriculture. Now, these worms are used in production of high quality compost called vermi-compost. It is rich in nutrients (N, P, K and Mg) and also in bacterial and actinomycetes population. The production of vermi-compost from an organic-waste by using composting earth-worms and its utilization in agriculture is one of the most economic ways in keeping the soil alive for sustainable crop production.

The verni-composting and vermi-culture is not a new technology as it has a long history; however, earlier, it was not adopted scientifically. But, now it becomes very essential to adopt Vermi-compost and Vermi-wash technology scientifically for economic point of view as well for stability and sustainability in agricultural production.

Vermi-wash

The liquid secretion collected from the body of earth-worm is called vermi-wash. It is found rich in plant nutrients and hormones. It can be sprayed on the crops for balanced development of crop plants and higher yield.

Benefits of vermi-compost and vermi-wash

Vermi-compost and vermi-wash play an important role in production of organic quinoa in the following manner:

- Both inputs improve quality of quinoa grains in terms of size, taste and keeping quality.
- Reduces cost of production.
- They improve physical, biological and chemical properties of the soil.
- Maintain balance of macro and micro-nutrients and reduce deficiencies of different nutrients in the soil resulting better yield of quinoa.
- Humus substances present in vermi-compost and vermi-wash help proliferation of soil microbes even in presence of chemical fertilizers.
- There is more stabilization of yields in quinoa and crop stands better under water and nutrients stress conditions.
- Due to better supply of micro-nutrients, the immune system of the quinoa plants improves considerably which results very less attack of insects-pests and diseases.
- Amelioration with chemical fertilizers reduces volatilization and leaching of nitrogen and enhances mineralization of phosphorus in the soil.
- A proper combination of BDS, bio-fertilizers and other inoculants such as rock-phosphate, blue green algae can give sustainable yields of quinoa at higher levels.
- It is good medium for good cultivation of quinoa.
- Vermi-compost is applied at the time of field preparation. About 10 to 15 q vermi-compost is recommended for quinoa cultivation for obtaining better yield. Vermi-wash is sprayed on the plants. 1 to 1.5 liters is sufficient for spray on the one acre quinoa crop.

C. Bioplantomin

Bioplantomin, also known as Super Plant Booster, is an organically certified and entirely safe product manufactured by 3D Industries Pvt. Ltd, New Delhi. This input is approved by Ministry of Agriculture, Government of India for organic farming and also certified organically under USDA, EU and Indian standards. It is a multi-functional (all-in-one) plant food which contains **(I) Essential nutrients** (primary, secondary and micro-nutrients), **(ii) Phyto-growth hormones** (similar to gibberellins, cytokinins and auxins in action), **(iii) Enzymes** and **(iv) Bio-active substances** (anti-fungal, anti-bacterial and anti-insect growth). It works as **Sanjivani** (plant booster) for quinoa plants by making them strong, vigorous and resistant to pest, frost and drought. It is used for foliar sprays, seed treatment and root dipping of quinoa seedlings before transplanting. All its constituents are derived from organic (living and plant) sources. It is recommended for commercial use in cultivation not only for quinoa but also for all agricultural / horticultural crops and forestry plants under different agro-climatic conditions and different type of soils for bumper harvest.

Composition

(a) **Primary nutrients (Minimum):** Nitrogen (N): 1.5 %, Phosphorus (P): 0.8 %, Potassium (K): 0.8%

(b) **Secondary nutrients (Minimum):** Calcium (Ca): 0.2 %, Magnesium (Mg): 0.5%, Sulphur (S): 0.5 %

(c) **Micro-nutrients (Minimum):** Iron (Fe): 25 ppm, Boron (Bo): 10 ppm, Molybdenum (Mo): 2 ppm, Zink (Zn): 15 ppm, Copper (Cu): 8 ppm, Manganese (Mg): 15 ppm, Chlorine (Cl): 30 ppm.

(d) **Growth promoting substances:** Various growth promoting substances like Cytokinin, Gibberellins and Auxins in action are present in bioplantomin which enhances growth and development of plants.

(e) **Other beneficial substances:**

S.No.	Compounds	Action
1.	Embaline	Retards bacterial growth and prevents occurrence of bacterial diseases
2.	Allicine	Retards fungal growth and prevents
3.	Azadirachtin	Retard insect growth and prevents attack of insects
4.	Enzymes (Protease inhibitor)	Forced insects to die due to blockage of digestive system
	Trypsion	Trace
	Cathepsin-D	Trace

Application benefits: Use of bioplantomin provides various benefits to the quinoa growers as mentioned below:

- More economic yield (20 to 40 %).
- Saving of seed input (up to 15 %) per unit area.
- Fertilizer saving up to 30 percent.
- More economic return per unit area.
- Low mortality of plants after transplanting.
- Improvement in seed germination.
- Excellent growth of yield attributes.
- Enhancement in health and strength of the plants.

- Increase in fertilizer use efficiency (60 to 70 %).
- Prevention of physiological disorders like low flowering, low fruit setting, seed dormancy, alternate bearing etc.
- Increase in resistance in plants against drought, frost and insect-pests and diseases.
- Increase in quality of crop produce like taste, nutritional status, colour, size, keeping quality etc.
- Prevention of nutritional deficiency.

Method of application: The use of Bioplantomin in quinoa cultivation is given in following table:

S.No.	Method of Application	Recommended concentration	Recommended Dose/acre
1	Seed treatment	2 ml/liter of water	10 liters water to soak 1 to 2 kg seed of one acre
2	Aerial spray	2 -3 ml/liter of water	100 to 120 liters water (200 – 250 ml) per spray in one acre

D. Harit Vardan

Harit Vardan is also an organically certified product of 3D Industries Pvt. Ltd, New Delhi. It is also approved by Ministry of Agriculture, Government of India for organic farming and certified organically under USDA, EU and Indian standards. All its constituents are derived from organic (living and plant) sources, thus, it is considered entirely safe and highly suitable for organic farming of quinoa. It is a soil amendment, engineered by a high-tech blending of Phyto-hydro-absorbent of high-grade performance, slow release bio-compounds, essential nutrients, plant growth stimulators (hormones) and insects-pest growth retarding bioactive substances. It is more effective than any single water absorbing Polymer or fertilizer with significant effect on growth, yield and quality of crop produce. It is recommended for quality production and enhancement of yield of quinoa under marginal environment.

Composition

S.No.	Ingredients	Amount
1.	Phyto-hydro-absorbents	85 %
2.	Primary nutrients (N,P,K)	9.5 - 12.5 %
3.	Secondary nutrients	
	Calcium	1.5 – 1.7 %
	Magnesium	1.5 – 1.6 %
	Sulphur	0.6 – 0.9 %

4.	Micro-nutrients	
	Iron (Fe)	25 – 45 ppm
	Molybdenum (Mo)	2 – 9 ppm
	Boron (B)	10 – 25 ppm
	Copper (Cu)	18 – 20 ppm
	Manganese (Mg)	15 – 22 ppm
	Zink (Zn)	14 – 28 ppm
5.	Slow release compounds	3 – 4 %
6.	Growth promoting substances (Similar in action to Auxins, Gibberellins and Cytokinins)	Trace
7.	Bio-active compounds Anti-fungal - Allycine, Anti-bacterial - Embaline, Insecticidal - Azaderectin, Anti-drought - Mycorrhizal, Anti-frost - Mycorrhizal, Protease inhibitor enzymes (Tripsin and Cathepsin-D)	Trace

Application methods

It is applied in the soil at 6 to 8" deep in root zone of quinoa plants. In nursery beds of quinoa, it is mixed in the soil 2–4 inch deep before seed sowing. When transplanting is followed in quinoa, it is applied in pits or holes before inserting plants. 5 Kg Harit Vardan is recommended per acre in case of quinoa.

Application benefits of Harit Vardan

- More economic yield (up to 40 %).
- Water saving up to 50 percent.
- Fertilizer Saving up to 80 percent.
- Decrease in mortality of plants.
- Improvement in seed germination.
- Improvement in physical properties of soil (Aeration, water holding capacity, soil texture, soil structure, porosity, field capacity etc.).
- Improvement in biological properties of soil.
- Increase in fertilizer use efficiency (60 to 70 %).
- Enhancement in water use efficiency (50 to 70 %).
- Increase in resistance in plants against soil pathogens, drought, frost, and insect-pests and diseases.
- Increase in quality of crop produce i.e., taste, nutritional status, color, size, keeping quality etc.
- Prevention of nutritional deficiency.

Bioplantomin and Harit Vardan, both are non-toxic (free from oral toxicity, dermal toxicity, eye irritation, cell toxicity), non-pollutant (free from carcinogenic substances), and non-carcinogenic, non-toxic to farm-animals, wild animals, bacteria, fish, birds, earth-worms, soil micro-organisms etc. Both products were tested under field trials of quinoa in different states during 2009–10 to 2013–14 by the Sarvati Devi Memorial Trust (pioneer Agency of quinoa development in India) under project entilted "Super foods for Nutritional Food Security and Doubling Farm-income (SNFDFI)" conducted in different states and these products were found very effective in respect of increasing yield (20 to 34%), quality and economic return of quinoa.

Due to encouraging results in quinoa cultivation, these products were included in growing protocol of quinoa in national quinoa development programme and supplied to the quinoa growers free of cost during two years i.e., 2014–15 and 2015–16. Both products presently being made available globally to quinoa growers by 3D Industries Pvt. Ltd, New Delhi since 2016–17 onward and strengthening their availability network in India and abroad for organic quinoa growers.

E. Farm Dhan (Nano-compost)

- It is nutrient rich manure which contain growth promoting compounds, soil fertility renewal substances, plant immunity enhancing and life saving bio-active substances.
- It works as "SANJIVANI' of plants and make them strong, vigorous, healthy and disease free.
- It ensure crop production in rainfed / un-irrigated, water scarcity, extremely drought, frost and un-favourable conditions.

Components in Farm Dhan

1	Primary Nutrients	N, P, K
2	Secondary Nutrients	Ca, Mg, S
3	Micro-Nutrients	Fe, Cu, Zn, Mn, Bo, Mo, Cl
4	Growth promoting compounds	Phyto-hormones like Gibberelins, Cytokinins Auxins etc.
5	Soil fertility renewal substances	Enzymes, humic acid, carbon, actinomycetes and other micro-organisms
6	Water absorbents	Phyto-absorbents
7	Life saving substances	Antibiotics, allycin

Comparision with convensional compost and Farm Dhan (Nano-compost)

S.No	Conventional composts	Farm Dhan (Nano-Compost)
1	Fails to deliver vital nutrients in shorter time	Deliver all nutrients in shorter time
2	Fails to augment water Use Efficiency	Augment water Use Efficiency up to 95%
3	Contain lot of weeds resulting problem of weed infestation	Free from weed seeds
4	Phyto-growth hormones are absent	Present
5	Heavy Metals present more than permissible limit	Free from such metals
6	Negligible impact on soil water holding capacity	Increases 60 to 70% water holding capacity
7	No effect on seed germination	Seed germination significantly increases
8	No effect on soil born pathogens	Prevent insect and pathogens growth up to 80%
9	No effect on biological resistance of plants	Augment biological resistance power of plants up to 80%

13.4 RECOMMENDED BIO-PESTICIDE INPUTS

Biopesticides are prefered over chemical pesticides due to following reasons:

- They do not leave hamful residues.
- They are target specific and do not destroy beneficial organisms.
- They promote the growth of natural enemies of pests, thus, reducing the need for future pesticide application.

The most effective, beneficial and recommended biopestices for plant protection in quinoa crop are as under:

A. Ultrabiogin (Liquid bio-pesticide)

- It is multi-action bio-pesticide designed to control vide range of insects and fungal, bacterial and viral infections.
- It is biodegradable and completely safe to human beings, farm animals, environment and crop produce.
- It is used in spray, seed treatment and root treatment to prevent insects multiplication and infestation of fungal, bacterial and viral diseases.

It can be sprayed on crop as per requirement. For spray, a solution is prepared by dissolving Ultrabiogin @ 4 - 5 ml / liter of water.

Components of Ultrabiogin biopesticide

S. No.	Compounds	Effect/Function
1	Azadrectin	Reduces insect growth
2	Phyto-thiophenes	Prevent growth of virus, namatode and insects
3	Retinoid	DNA damage in cells
4	Phyto-acdysones	Prevent insect survival

B. Green power (Solid biopesticide for soil application)

- It is multi-functional soil conditioner / Bio-pesticide.
- It prevents / suppress growth of soil insects, nematodes, rodents, mites, wide range of soil born pathogens (fungi and bacteria) and prevent damage caused by them.
- It is biodegradable and completely safe to human beings, farm animals, environment and crop produce.
- It is applied in the soil at the time of seed sowing / transplanting of quinoa seedlings.

Components of Green power:

S.No.	Compounds	Effect / Function
1	Nutrients (NPK)	Role in physiological activities and photosynthesis
2	Growth promoters	Growth enhancement
3	Azadrectin	Reduces insect growth
4	Phyto-thiophenes	Prevent viral, nematode and insect growth
5	Retinoid	DNA damage in cells
6	Phytoecdysones	Prevent insect survival

Mode of action of Ultrabiogin and Green Power biopesticides:

- It destroy eggs of insects and larva and not allow insects to multiply in soil.
- It damages DNA and other compounds of mitochondria of insect cells.
- It interferes with nervous system of insects and damage cell membrane resulting leakage of cell's fluid and cause ultimately death of insects.
- It alters hormones of insects resulting failure to breed, eat and fly.
- Crops are safe due to direct killing, repelling and death of insects due to starvation.

Comparision with chemical pesticides

Chemical pesticides	Ultrabiogin and Green Power biopesticides
Pose a serious threat to environment, human, birds and animals	Completely safe
Non bio-degradable	Bio-degradable
Harmful nuclides (cobalt, radium, uranium, dioxins etc.) are present in considerable amount	Not present
Highly toxic to cardio-vascular, neurological, reproductive, immuno-logical endocrine, respiratory systems of human body	Non toxic to all human systems

C. Bio-control agents

Biological control agents are naturally occuring enemies of pests. Biological control uses these natural enemies in the direct manner to control pest population and follow three strategies:

- Classical biological control, where ecologically adapted natural enemies are introduced from the area of origin of pest to the target area.
- Conservation of natural enemies present in the ecosystem using cultural practices or habitat management that enhance their activity.
- Artificial augmentation of local natural enemy population.

Biological control has the advantage of eliminating the need to use chemicals and low in cost. In addition, it is self-sustaining, may not require inputs from farmers and is safe for the environment and human health. Biological control agents seem to be particularly effective in controlling the population of exotic or alien pests which tend to proliferate in the absense of natural enemies in a new ecosystem.

The description of best biological control agents are given below:

- **Arbuscular mycorrhiza (AM fungi):** It is believed that the development of the arbuscular micorrizal symbiosis played a crucial role in the initial colonization in the soil by plants and in evolution of the vascular plants. Under programme entitled "Extension Research Systems for Quinoa Technology Transfer", Quinoa performed better and gave increased yield upto 8% when soil is well colonized by AM fungi. Management of this fungi is an important for organic and low input agriculture systems where soil phosphorus is low. This fungi improve the quality of soil and the productivity of land. AM fungi help quinoa plants to capture nutrients, such as phosphorus, sulphur, nitrogen and micro-nutrients from the soil. Availability for pure cultures of Arbuscular Mycorrhiza fungi is an impediment in large scale production despite the fact that beneficial effects of AM fungal inoculation to plants have been repeatedly shown under experimental conditions in the laboratory especially in conjunction with other nitrogen fixture.

It is an arbuscular mycorrhizal fungus (Plural - mycorrhizea or mycorrhizas) is a type of mycorrhiza in which the fungus penetrates the cortical cells of the roots of a vascular plant. Arbuscular mycorrhiza are characterized by the formation of unique structures, arbuscules and vesicles by fungi of the phylum Glomeromycota. The transfer of nutrients mainly phosphorus, zinc and sulphur from the soil *milleu* to the cells of the root cortex is mediated by intracellular obligate fungal endo-symbionts of the genera *Glomus, Gigaspora, Acaulospora, Sclerocysts* and *Endogone* which possess vesicles for storage of nutrients and arbuscles for funneling these nutrients into the root system. By far, the commonest genus appears to be *Glomus*, which has several species distributed in soil.

- **Bacillus thuringiensis:** The biopesticide *Bacillus thuringiensis* (Bt), among the most widely known and researched biocontrol agents, is a microbe that produces a special protein that is active against a very narrow spectrum of insects. Different strains of Bt are used against different pests.

The neem tree and its various products are among the important botanical pesticides. Neem contains several chemicals that affect the reproductive and digestive process of a number of important pests. Neem also acts as a repllent and anti-feedant and its oil is effective against leaf folders, aphids, and bollworms. About 200 species of insects are known to be controlled by neem.

Chapter

14

INSECT-PESTS AND THEIR CONTROL MEASURES

14.1 INTRODUCTION

A wide range of insect-pests are reported in the Andean region, Bolivia and Peru which damage quinoa crop during seed germination to crop harvest and grain storage. During the period of vegetative growth in Andean region, the quinoa crop is affected by a wide range of insects and loss due to insect-pests recorded from 8 to 40 percent. In other regions, insect-pests are not considered as a serious problem due to smaller area under quinoa. The situation is similar in new quinoa growing areas around the world. In conventional quinoa farming, various kind of chemical pesticides are used like other crops to control pests, while in organic farming, an integrated plant protection approach is needed that relies on various agronomic practices and crop inputs that meet organic standards.

In Andean region, the problem occurs in the cropping season due to miners and borers eating leaves, stem, roots and grains, chewing and sucking insects on foliage and stem cutters, birds and rodents feeding on mature grains. Pests also attack quinoa grains during storage. In new areas, other than Andean region, Bolivia and Peru, the studies on insect-pests are very less. This chapter focuses on insect-pests noticed and cause major damage on quinoa crop at different regions of the world.

14.2 INSECT-PEST OF QUINOA

The most economically important and endemic pests identified in different quinoa growing regions are quinoa moths (*Eurysacca melanocampta, Meyrick*) and the ticona complex (*Copitarsia turbata, Feltia sp, Heliothis titicaquensis, Spodoptera sp).* The damaging insects of quinoa found elsewhere in the world are polyphagous pests constituting a wide range of potential pests if quiona is grown as a crop in those regions. Other major pests include a group of cut-worms (Noctuidae). The details of most serious and damaging insects of quinoa crop are given below.

- **The Quinoa Moth:** The quinoa moth is considered as a serious pest of quinoa in most of the quinoa producing countries. The quinoa moth belongs to *Eurysacca* genus, Gelechiidae family and Lepidoptera order. More than 20 *Eurysacca* species have been recognized, but, *Eurysacca quinoae* and *Eurysacca melanocampta* are important, abundant, wide spread and most destructive species causing more economic losses of quinoa. Its larvae damage in the stage of developing flowers and grains maturation. The worst damage by moth is observed between the stages of grain development and physiological maturity, when the larvae feed mainly on tender leaves if they are in their first stages or immature and mature grains in later stages. The harm done by the quinoa moth occurs at two levels: indirect and direct larval damage to the plant. With regards to indirect damage, the photosynthetic area of the plant is reduced and first generation larvae feed on the parenchyma of the leaves, roll leaves and tender shoots and destroy developing inflorescences. Second generation larvae destroy the developed inflorescences, milk and dough stage grains and mature grains, thereby reducing the quality and quantity of the grain yield by 15–60%.

Eurysacca melanocampta is more destructive insect which may cause yield losses from 15 to 18% or more in dry environment. Field observations indicate that sweet / white varieties of quinoa with low / nil saponin content, are more susceptible to the attack of this insect than bitter ones. They can be recognized by their wing patterns. *Eurysacca melanocampta* has a longitudinal dark band on the wings and two dark spots surrounded by light scales.

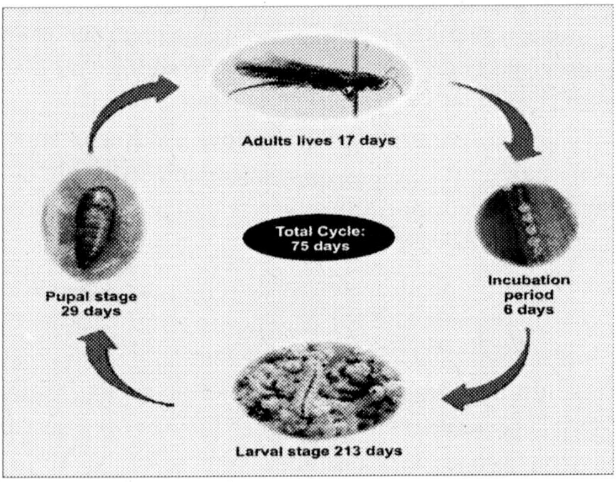

Fig. 25: *Life cycle of Eurysacca melanocampta Moth*

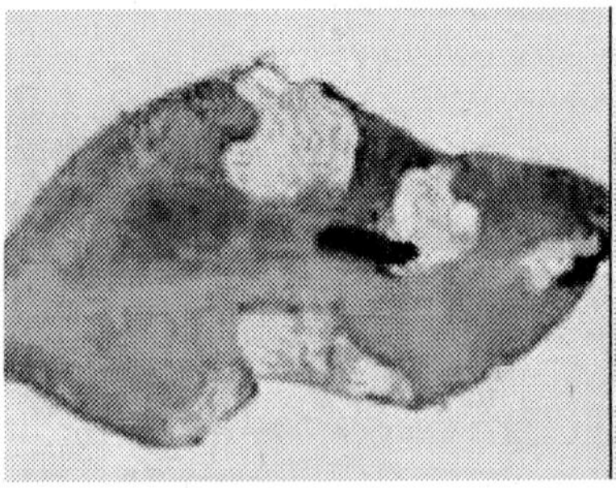

Fig. 26: *Leaf damage by Eurysacca melanocampta*

Fig. 27: *Panicle damage by Eurysacca melanocampta*

Eurysacca quinoae has two dark spots on the wings as well as some small dark spots at apex of the wing, but it does not have a dark longitudinal band. It has been reported in Peru and Bolivia as feeding only on quinoa.

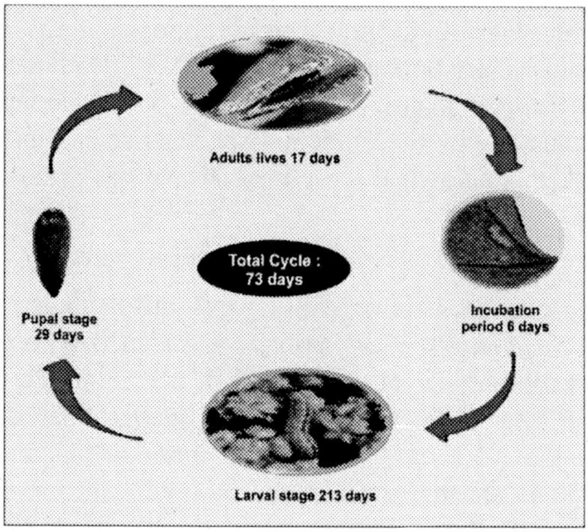

Fig. 28: *Life cycle of Eurysacca quinoae moth*

Both kinds of insects has four life stages i.e., eggs, larva, pupa and adult. The adult of moth feeds on flower nectar and the damage is on panicle. It is a small about 8 to 9 mm in length with a wing span of 14 to 16 mm. The females of this moth lay their eggs (sub-globulose in shape, smooth, creamy white in colour) on the leaves in the pre-flowering state when the grains are already developed. The eggs are 0.6 mm long and 0.3 mm wide. Five (I to V) instars exist pale green to yellow with several narrow red-maroon longitudinal bands. Pupae are dark maroon.

The larvae of first stages feed on the leaves and subsequently on the developing and maturing grains. Mating occurs over quinoa plants or over plants nearby. Larvae attack the quinoa crop in two generations; the first generation (November and December) consumes and destroys the leaves and forming inflorescences, stick young leaves together and roll them up. The second generation (March–May) of larvae attack plants in the maturation phase, feeding on the forming and maturing grains inside the panicles. The second generation causes the most economic damage to the quinoa crop.

- **The noctuid / ticona complex:** The noctuid complex refers to a group of insects belonging to the *Helicoverpa quinoa, Capitarsia incommoda* and *Agrotis ipsilon* genera whose larvae cause serious damage to quinoa crops, especially in Bolivia, Peru, Chile, Columbia, Ecuador, Argentina and Peru. The adult is a nocturnal moth or butterfly that is attracted by the light. The larval instars cut emerging plants of quinoa near the ground.

  Adult insects of this group do not damage quinoa crops because they feed only on flower nectors and sweet secretions from plants. Hatched larvae mine the developing inflorescence, causing branching in the quinoa plant on which small penicles form.

During the plant development stage and when the larvae are bigger, they feed as defoliators. During the flowering and crop maturity stage, larvae cause major damage as they bore into the panicle rachis leading it to break off which results in defoliated plants. They also feed on developing grains. The most significant damage by these noctuid species occurs during the grain's milk stage or dough stage when these pests behave as grain feeders and have a direct impact on yield.

The most destructive insects of this group are descrived here:

(i) **Helicoverpa quinoa:** It is very difficult to differentiate *Helicoverpa quinoa* species from *Helicoverpa gelotopoeon* and *Helicoverpa titicacae* species on the basis of morphological traits alone. It is also not clear that *Helicoverpa gelotopoeon* and *Helicoverpa titicacae* species that infests many crops throughout the world, are also quinoa pests or not. However, due to its polyphagous nature, these could pose a significant problem in new quinoa growing areas. *Helicoverpa quinoa* is the most serious pest of quinoa grown in Andean region due to sizeable yield losses upto 20%. It has a very particular life cycle. The incubation period lasts 5 ± 1 days, the larval stage 26 ± 3 days, the pre-pupal stage 9 ± 1 Day and pupal sage 175 ± 29, while the adults lives 6 to 10 days.

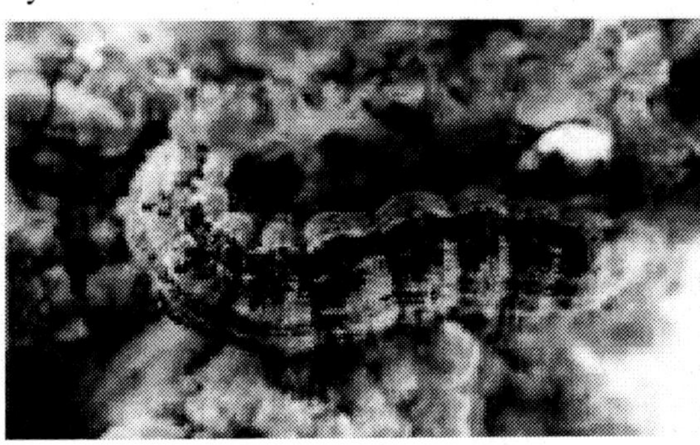

Fig. 29: *Helicoverpa quinoa larvae feeding on quinoa grains*

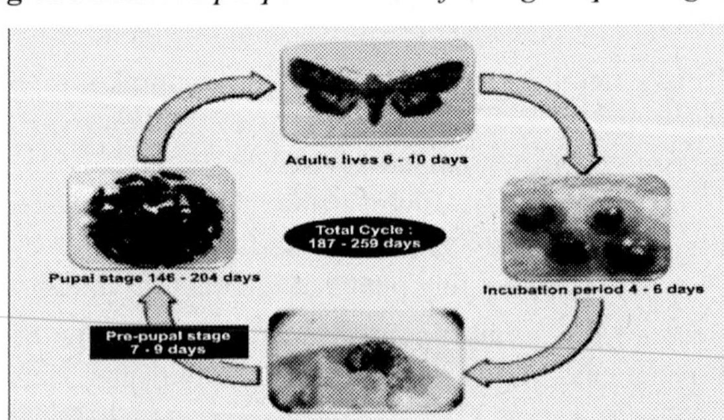

Fig. 30: *Life cycle of Helicoverpa quinoa*

- **Copitarsia incommoda:** The polyphagous larvae is found from Maxico to Chile and its polyphagous beheviour cause considerable economic losses to many crops. This species is reported in Bolivia and Peru as one of the principal quinoa's pests, especially in the area around Lake Titicaca where this pest and the quinoa moth cause economic losses of around 30 percent. The polyphagous behaviour of this insect, is that it feeds on various plant species and the fact that it is present in many areas around the world, make it a potentially highly destructive pest anywhere quinoa cultivation is introduced and developed.

  Egg incubation averages 5.5 days, the larval stage 26.13 days, the pre-pupal stage 3.09 days and the pupal stage 16.3 days, while the adults live 19.85 days with the species completing its life cycle in 70.87 days.

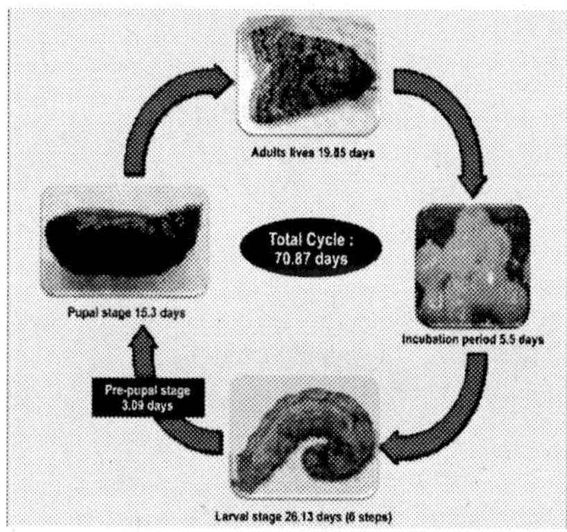

Fig. 31: *Life cycle of Capitarsia incommoda*

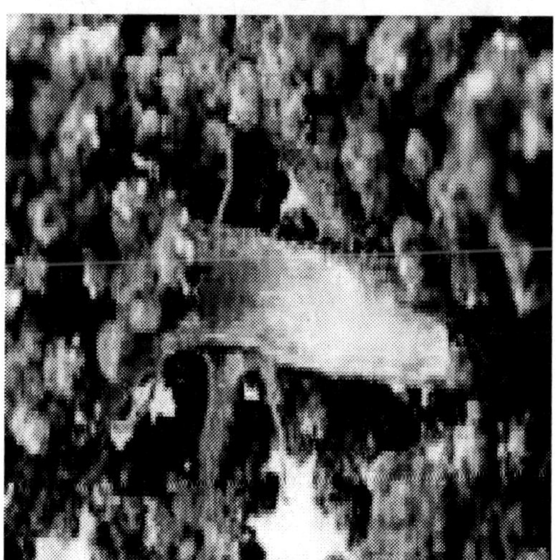

Fig. 32: *Damage to the panicle rachis caused by Copitarsia incommoda*

- **Agrotis ipsilon:** *Agrotis ipsilon* larvae, commonly known as cut-worm, is found throughout world, particularly in Andean region where they are considered damaging pests of various crops. Larvae live below the soil where they build protective cell. At dusk and night, they come out to feed on seedling stems, leaves and roots. The larvae in the first stages of development are mainly defoliators, becoming cutters in later stages. They mostly spent summer as larvae. Pupation occurs in the same underground cell. Adults can emerge year round, but, do so primarily in autumn. This insect can over winter as larva or pupa. Its larvae cut the seedling stem at the ground, but, can also feed as defoliators.

The eggs hatch in 3.83 ± 0.17 days, the larval stage lats 20.6 ± 0.93 days, the pre-pupal stage 2.22 ± 0.21 days, the pupal stag 12.51 ± 0.36 days and adults live 18.91+-3.36 days for a complete life of 57.96 ± 5.03 days.

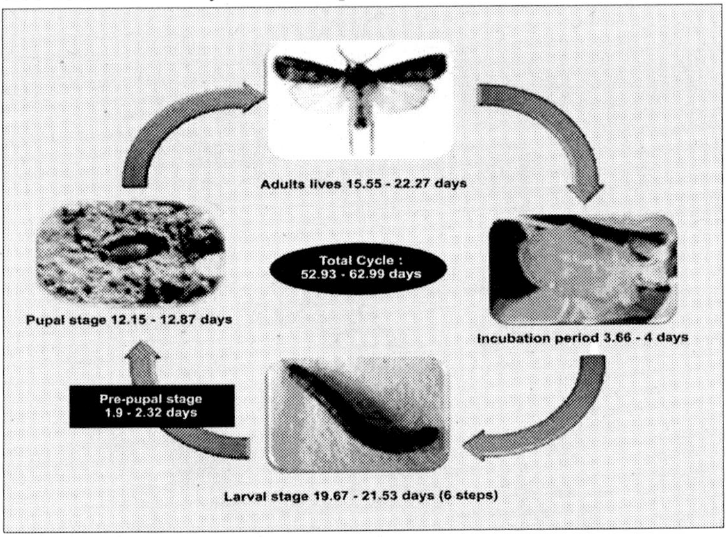

Fig. 33: *Life cycle of Agrotis ipsilon*

14.3 CONTROL MEASURES OF INSECT-PESTS

- ***Biological method:*** Biological control methods are recommended in quinoa production to maintain quality of grains. Insects of Noctuid group can be controlled by a range of parasitoid species as well as predators upto 45 percent in the field. Studies on biological control method of quinoa insects is mostly lacking.
- ***Using bio-insecticides:*** Ecofriendly bio-insecticides can be used in organic quiona production as they are bio-degradable and do not harm the environment. Bio-insecticides and eco-friendly pesticides recommended for organic quinoa production have been discussed in Chapter - 13.

  Quinoa plant display a large amount of oxalates on the surface of leaves, stems and panicles, reducing product adherence, it is very important to use an adjuvant which helps sticking. Vegetable oil spray acts as dispersal agent, improves coverage and prevents the formation of large drops. The application of vegetable oil spray adjuvant enhances product efficiency.

Table 14: *Eco-frinedly Bio-pesticides recommended for insect control in quinoa*

Bio-insecticides	Dose	% Efficacy
Bacillus thuringiensis	90 gm per 20 liters	63
Ultrabiogin (Liquid)	3 ml per liter of water	93.5
Green Power (Soild)	2 Kg per Acre	95
Lime sulphur	500 cc	35

- **Using Chemical insecticides:** Chemical insecticides have advantage of being quick-acting, effective and economical. However, they have long-term negative effect on beneficial insects, environment and farmer's health. They can be used in conventional quinoa production only. In non-organic production, spray of Melathion insecticide dissolved @ 2 ml per litre of water can be sprayed as per need to control the wide range of insects of quinoa.

  Future research should focus on the main pests in order to reveal basic information on interactions with the host plants. Population carry over from one growth season to another and the role of environmental factors on insect development and population size needs to be studied. Cultural practices and host plant resistance are important component of integrated pest management (IPM).

- **Integrated Pest Managemnt:** The economic and ecological impact of pesticides abuse has prompted attempts to reduce pesticides use and led to an increased investment in alternative technologies and products, a reconsideration of traditional methods of pest management, and a more comprehensive look at the ecological context of pests including an increased interest in IPM.

  The techniques used in crop protection in IPM may include crop rotation, light traps, pheromones, sanitation, mulching, tillage, intercropping etc. The options used by the farmers depend on the local context-agro-ecological needs, availability and affordability of the various alternatives.

The following methods are important and effective:

- **Crop rotation:** Crop rotation and inter cropping has been recommended to break the continuity of the food chain for quinoa insects and also to break the pest life cycle. Ploughing of field is beneficial before sowing of quinoa seed as pupae of harmful insects exposes to temperature, birds and other predators.
- **Using Light traps:** Light traps are advised to attract adult moths to capture and kill them. The basic design of light trap is a bright light sources and a capture mechanism containing water and small amount of detergent to reduce the surface tension and prevent the insects from escaping. Light trap captures a wide range of insect species.
- **Using Pheromones:** Pheromone traps are also useful method of insect control in quinoa. Sex pheromones are glandular secretions from the males that cause specific attraction reactions in males of the same species. Such pheromones can be used to monitor pests, to control adult insects or to disrupt mating. Pheromones are not harmful to the environment, accepted in organic farming and are effective for atleast three months.

Chapter

15

DISEASES AND THEIR CONTROL MEASURES

15.1 INTRODUCTION

Quinoa is infected by a variety of pathogens in Andean region, Bolivia and Peru in which damping off (*Sclerotium rolfsii*), downy mildew (*Peronospora farinose*), brown stalk rot (*Phoma exigua* var. *foveata*), leaf spot (*Ascochyta hyalospora, Ascochyta chenopodii and Passalora dubia*), bacterial spot, grey mold (*Botrytis cinere*a), seed rot and damping off (*Scerotium rolfsii, pythium zingiberum*) and bacterial blight (*Pseudomonas* sp.), nematodes and viruses are most destructive diseases of quinoa crop. Except downy mildew, other diseases are less widespread and of less economic importance, but are still considered potential production constraints particularly when quinoa is introduced in areas outside its traditional growing regions. Combined with the effects of climate change, they could become more serious. The incidence and severity vary according to the variety, phonological stage, and environmental conditions. Furthermore, in other countries around the world with different ecological and environmental characteristics, it is probable that new diseases will emerge.

15.2 IMPORTANT DISEASES OF QUINOA

Downy mildew

It is most significant, well-known, serious, widespread and primary disease of quinoa on a global scale than other ones. It was first reported in Peru in 1947 and has since been found in numerous countries across the globe. It is known to cause yield reduction from 33 to 58%. The highest damage due to Downy mildew occurs during the flower-bud initiation stage and thereafter, incidence gradually decreased. Quinoa showed high level of resistance towards downy mildew in North-Indian conditions. It is caused by the fungus *Peronospora variabilis* (formerly called *Peronospora farinose*), an obligate biotrophic parasite from the Oomycetes group. It is easily transmitted via wind and rain. The fungus has two types of reproduction, i.e., Asexual (ovoid spores with direct germination) and sexual (via oospores formation / survival structure in the absence of host plant). When conditions are favourable, oospores germinate and produce spores. This disease is essentially spread over short distances through spores and over longer distances through oospores.

The incidence and severity of mildew is depend on the variety's degree of resistance, crop management approach, phonological stage of the plant when attacked and environmental conditions. If attack is occurs during initial development stage of plants, the panicle formation atrophies (slow down in growth) and grain filling as well as size of grain are seviourly affected and the entire crop can be lost. In resistant varieties, yield losses can be in the range of 15 to 30%. If the climate is favourable during dough stage, the grains may turn black. In resistant varieties, grain size does not affected. When the disease is appears after flowering, it can be confused with the plant's natural senescence (general yellowing), but, does not cause considerable losses.

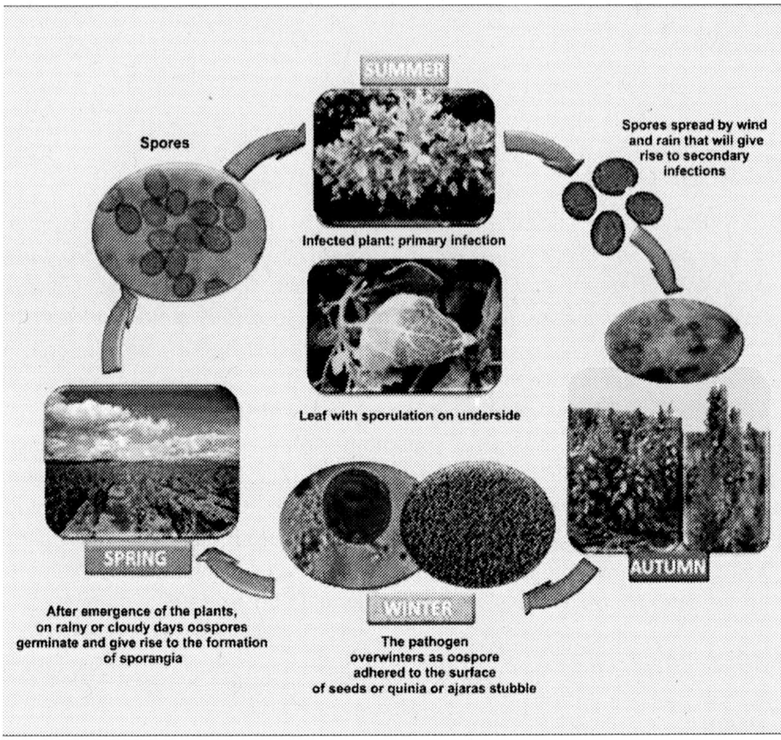

Fig. 34: *Quinoa downy mildew cycle*

The disease is less sensitive to temperature and can develop at temperatures between 0 to 25°C. Wet area or period of relative humidity more than 80% and temperatures between 18 to 22° C promote spores formation and fungus growth resulting more spread of the disease. Cloudy weather without rain also promote the appearance of this disease. In dry regions, mildew is not a serious problem. Oospores form when the crop is in senescence or when conditions become favourable to the pathogen.

Fig. 35: *Typical spots caused by Peronospora variabilis*

Symptoms

The disease is primarly affects foliage and reduces photosynthetic areas of plants (appearance of chlorotic or necrotic spots on the leaves), causing partial or total leaf loss, atrophied plant development, reduced panicle size and lower yield (small or defective grains). The symptoms may also appear on stems, branches, inflorescence and grains. Initially, symptoms appear on the leaves as small, irregular spots that may be chlorotic, yellow, pink, red, orange or grey, depending on the plant colour. As the disease progresses, these spots begin to merge (coalesce), the leaves become chlorotic and eventually fall. When conditions are very conducive to disease development, all the leaves may become infected and fall off, halting plant growth. The spores of fungus are formed on lower side of leaves as greyish, while in resistant varieties, fungus growth may or may not appear. At plant maturity, the oospores adhere to the out side of the grains. Quinoa wild relatives are found in nearly all agricultural areas around the globe, they could be an important source of infection in new quinoa growing areas.

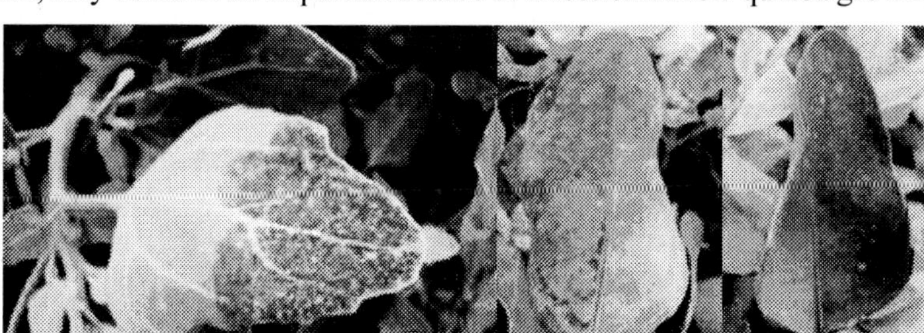

Fig. 36: *Spore formation on top and under side of leaf*

15.3 CONTROL MEASURES OF DISEASES

Under conventional farming of quinoa, the newest generation of fungicides can provide satisfactory disease control if they applied early as a preventive measures. For organic farming, there are a number of considerations to make to ensure that acceptable levels of severity are maintained. They incluse the use of resistant varieties like Sarvati (S-7), early planting, low planting density, and bio-fungicides approved for organic farming. Downy mildew can be treated with spray of liquid extract of garlic *(Allium sativum),* when disease is identified.

Viruses, causing chlorotic mosaic and false nodule caused by Nematode *Nacobbus* sp. and *Thecavermiculatus* sp. are also known to infect quinoa plants, but reports of significant damage are lacking.

15.4 OTHER PESTS OF QUINOA

Apart from insects, birds also cause a major loss to sweet quinoa growers primarily in the inflorescence stage. In bitter species of quinoa, the damage occurs very minor as such species contain a chemical substance in grains in the form of saponins. Rodents are also feed on quinoa grains, but are not commonly encountered in quinoa field. They seem to be a more serious as post-harvest pests when grains are being dried and in storage.

Chapter

16

POST-HARVEST HANDLING, PROCESSING AND STORAGE

16.1 POST-HARVEST OPERATIONS

This activity through which the grain is obtained includes drying or stacking, threshing, venting and storage.

Drying or stacking

After harvesting, it is important to dry crop to avoid occurance of mold in storage. It can be left on trays in the hot sun or placed near an indoor heat source. Stir occasionally until it is as dry as possible. Grains should be thoroughly dried before storage.

Threshing

This consists in the separation of grains from the panicle. There are several different methods of threshing i.e., manual, semi-mechanical, mechanical and direct threshing.

- **Manual threshing:** It is one of the most difficult tasks in quinoa production, and is practiced in places inaccessible to vehicles. For this method, it is necessary to prepare in advance the 'Takta' which consists of a platform of clay, water and "jipi" (or some farmers use canvas and Polythine sheet instead) upon which the dried plants are threshed. A stick is used to beat the plants and the resulting plant material is then coarsely sifted and then winnowed to obtain the grain. The yield for a day´s work is 1.5 quintals / day.
- **Semi–Mechanical threshing:** In the Southern Altiplano the use of vehicles (tractors, trucks, etc.) has been adopted to carry out this method of threshing. A tarpaulin is extended on the ground and on top the dried plants are placed in two parallel, longitudinal piles over which the wheels of the motor pass. The panicles are placed towards the inside of the two rows so that in multiple passes the vehicle is able to separate the grains. This method processes about 10 q in 50 to 60 minutes. The difficulty is in the following sifting and winnowing which are performed manually.
- **Mechanical threshing:** In the Southern Altiplano, threshers such as the "Vencedora" and "Alvan Blash" are used, which after adjusting their hulling and sieving systems achieve yields of 10 quintals / hour and 7 q / hour, respectively. In both cases the cleaning is 95%. The "Herrandina" thresher has also been tried with a yield of 123 kg / hour. The formers of Maharashtra state tried threshers and they cleaned quinoa grains successfully.

Winnowing

This work involves the separation of perigonium and plant waste from the commercial grain. There are three methods of quinoa winnowing i.e., traditional, improved manual and mechanical.

- **Traditional Winnowing:** It is carried out manually on a blanket or traditional textile with the help of a small plate, this method of winnowing is dependent on the presence of moderate winds. The average output is 4 q. / day
- **Improved Manual Winnowing:** It performed with a machine or prototype which has an air flow regulator and separates the quinoa grain from the 'Jipi', 'Chiñi' and plant waste. This machine yields 6 q / hour.
- **Machanical Winnowing:** Mechanical winnowing is in use for the last six years, yielding 5 to 8 q/ hour depending mainly on the amount of 'Jipi' which the material to be winnowed contains.

16.2 PROCESSING

Quinoa grains are not uniform in size after harvesting and winnowing. Grain size varies between 1.4 and 2 mm in diameter and the grain contains impurities including chaff residue, twigs, leaves and small stones as well as broken, damaged, coloured, covered and unripe seeds. Quinoa is processed to obtain grains that meet quality standards in terms of size, impurities and satisfy bromatological and microbiological requirements. The grains, therefore, have to undergo a series of processes including preliminary sorting and removal of impurities; saponin removal, which is normally carried out using both hulling (dry method) and washing (wet method); drying; sorting by size; separating of different coloured grains and removal of residual impurities.

Preliminary sorting and removal of impurities

Once the quinoa has been purchased, it is stored in 100 kg bags of plastic or other materials in facilities that are capable of processing hundreds of tonnes of grain each month. Some large processing plants use metal silos, to avoid rodents and moths.

Before being transported to the processing plant, generally in 100-kg bags made of polypropylene or other materials, the initial product is sorted using simple sieves made of a plate perforated with 3 mm diameter openings and a woven mesh with a spacing of 1.2 mm between the threads. The processing speed is 100 kg every 2–3 minutes. The machine runs on a 1.5 hp motor.

The sorting process generates five products:

- Particulate matter (mainly dust and saponins).
- Light, coarse impurities (twigs, leaves).
- First grade grain (grain with a diameter of > 2.2 mm) (90–95%).
- Second grade grain (grain with a diameter of < 2.2 mm).
- Heavy impurities (stones).

Particulate matter is discharged into the atmosphere, impurities are discarded and second grade quinoa is either returned to the farmer or purchased at a lower price at the same time as the first grade. Both products are weighed on a scale.

Some processing companies are equipped with the CIFEMA grain sorter or with similar prototypes that sort the grain by size using two sets of different-sized, interchangeable sieves, which can also be used to sort different varieties of quinoa. The sorter runs on a 5 hp diesel engine and has a processing capacity of 700–1000 kg/hour. The sieves measure 60 × 100 cm, and are equipped with a mesh with 2 and 1 mm openings.

Saponin removal

Saponin present in the pericarp of quinoa seeds is a toxic glycoside. The quinoa grains with saponin or unwashed quinoa gives a bitter flavour and has anti-nutritional properties. This toxic substance can easily be identified by production of a soapy lather when placed in water and also by placing seeds in pure alcohol. Once saponin is removed, protein quality is unaffected.

There are apparently two types of saponin:

- A neutral saponin group (found in white quinoas) that can be used commercially in the production of pharmaceutical steroids.
- A more common type prevalent in the yellow quinoa cultivars which is used in soaps, detergents, beer production, fire extinguishers, photography, shampoos, cosmetics, and pharmaceuticals (synthetic hormones).

The process of removing saponins is one of the most important stages in grain processing and in recent years. Various appropriate technologies have been developed for removing saponins to levels within the acceptable limits, without affecting the grain's nutritional properties.

Processing quinoa to remove saponin can be done by following two methods:

- **Water washing:** Saponin is traditionally removed simply by washing. Washing of quinoa grains is a traditional method used in Andean Region. A certain quantity of grain (5–10 kg) is placed in 50-kg bags made of cloth or synthetic materials. The bag is then submerged in the water of a river and/or a brook and, held at both ends, it is moved up and down so that the grains rub together. The water helps the saponins to dissolve and they are washed downstream. The movement is repeated until there is no longer any foam in the water. The grains are then dried on zinc sheets laid outside.

  In this method, some farmers or housewives entails a successive washing of the grains using friction by hand or a stone to eliminate the episperm. Commercially, this method is probably not a viable option at present due to high cost of foam breakdown and drying grains and pollution of water bodies. The stone mill remove 33 to 40 percent of the seed as bran fraction and remaining saponin bran is left on the grain.

  In the National Agrarian University of La Molina (UNALM), Lima, Peru, a cleaning method for quinoa was designed by applying 30 minute soaking of grains in water and 20 minutes agitation at 70° C. This process can reduce saponins satisfactorily down to 0.04 –0.25%. First, the seeds are mechanically abraded against rough walls. The grain is then rubbed against sieves to separate the next layer, and the residues and saponin dust are eliminated.

- **Mechanically via abrasion:** In this method, de-hulling is involves by pearling the grain to remove the pericarp as bran. High saponin cultivars require more abrasion than low saponin types. In general, mechanical / dry methods are economical, simple and do not lead to contamination, but they are relatively inefficient, only eliminating 80 percent of the saponin. The extremely high loss of grain via mechanical milling indicates an area of research needs more work.
- **Modern system of saponin removal:** For many years, quinoa processing companies used or adapted machines, equipment and technology initially developed for processing rice, wheat, soybean and sorghum. The low volumes of production, compared with these other crops and the existence of only a small number of milling companies globally, provided little incentive for developing specific machines, equipment and technology for this sector.

In the last 10 years, however, quinoa has experienced a quiet boom: once a product consumed solely by the farmers growing it in the Altiplano and Inter-Andean valleys, it has become a global, high commercial value crop cultivated in extensive areas, not only in the countries where it originated, but in others where it has been introduced.

This phenomenon is mainly due to the increased demand for gluten-free cereals from the 0.4 percent of the world population that suffer from celiac disease; the increased demand for high quality, affordable organic products; and the implementation of efficient food programmes in various countries by organizations such as FAO.

Quinoa processing companies mostly use the combined method to remove saponins and comply with the established market quality standards. Nevertheless, the process has always presented major difficulties with regard to removal of saponins and impurities and concerning grain moisture content. There are currently 62 processing plants in Bolivia, comprising 16 percent artisanal processors, 27 percent semi-industrial and 57 percent industrial companies. Of the industrial processing plants, 40 percent are found in Oruro, 25 percent in La Paz and 35 percent in Potosí, Cochabam-ba and Chuquisaca. The technologies used range from artisanal technologies to very complex and sophisticated processes.

One of the most significant industrial contributions has been the technology developed by the Sustainable Technologies Promotion Centre (CPTS), which uses the physical properties of the seed episperm. The grain undergoes a cleaning process to remove impurities in a preliminary sorter, followed by saponin removal in a huller with dual compartments: i) the hulling system, and ii) the particle extraction and collection system.

16.3 LOSSES DURING QUINOA PROCESSING

Processing losses include quantitative, qualitative and economic reductions of the product as mentioned below.

- **Weight Loss:** Weight loss is a reduction in the quantity of the product due to animal damage or shattering.

Table 15: *Post-harvest losses in quinoa*

Phase	Losses (%)		
	Minimum	Maximum	Average
Collection	1	3	2
On-farm transportation	2	7	4.5
Threshing	2	6	4
Drying	1	5	3
Storage	2	6	4
Processing	2	10	6
Total	10	37	23.5

- **Quality Loss:** Loss of quality causes decreases in market value of the product. This includes the following:
- **Losses in physical traits:** The physical characteristics usually considered in evaluating the quality are seed shape and size; moisture content; presence of foreign matter; germinated, broken, or damaged seeds; stones; dust; leaf residues; glass or metal fragments; animal hair or excrements; and level of infestation by insects or micro organisms.
- **Alteration of nutritional qualities:** A reduction of vitamin, protein, or lipid content in quinoa may occur over time.
- **Losses due to reduction of germination rate:** The seeds should be able to germinate at a high rate.

16.4 STORAGE

Quinoa grains are perishable, however, shelf-life of grains can be increased by safe storage methods. To ensure optimum freshness, shelf-life and to maintain quality of grains, storage should be carried out in a clean, cool, dark, dry and ventilated environment at dark place. Grains may be refrigerated in an air tight container to keep out moisture and other contaminants. For long storage, quinoa can be stored in oxygen free freezers. Cooked quinoa can also be freezed well in air tight freezer containers. An important point to remember is that whole grain does not keep as long as refind grains because germ portion of the kernels can cause the grain to become rancid over a time.

It is recommended that the grains are stored in sacks made of woven llama wool, or new or good condition polypropylene bags. The filled bags should be stacked properly on a wooden pallet. Quinoa grains and seeds are stored in air-tight containers in a cool dry place. Quinoa grains can be stored safely from 2 to 3 years.

For keeping grains fresh at homes, grains should be kept sealed in containers with tight fitting lids at cool, dry and dark places. A sealed container is important to maintain freshness, and reducing possibility of infestations. Refrigerator can be a good place to store grains, especially in warm weather, however, using a sealed container is even more important in this case in order to prevent the grain from absorbing moisture and odors and flavours from other foods.

Fig. 37: *Preliminary sorter*

Fig. 38: *Machine of quinoa processing made in India*

Chapter

17

USES OF QUINOA AND ITS VALUE ADDED PRODUCTS

Quinoa is marketed in domestic and international markets in different forms i.e., processed grains and value added products. Considering its high nutritional and therapeutic values, quinoa is used in preparation of different kind of food and non-food products. Food products prepared from quinoa are considered as highly nutritious and healthy for children, pregnant women, celiac, women experiencing menopause, the elderly, convalescents, athletes, vegetarians, diabetics, very stressed people and adults in general.

17.1 USES OF QUINOA

There are two type of uses of quinoa as described hereunder.

Food uses: Quinoa is used mainly as nutritive food, traditionally like rice. It is also used in making breads, pulaw, khichari, Daliya, Bhature, masala dhosa, idali etc. Now a days, several processed or semi-processed value-added food-products are available throughout the world. Such products are highly nutritive and healthy with positive impact on digestive system. They are ready-to-eat foods generally used in breakfast. Such products include puffed grains, pasta, cakes, flakes, dry noodle, baby foods, drinks (soups, alcoholic beverage), mixed grain dishes, cookies (biscuits), namkeen bhujiya, quinoa milk powder etc. The fresh / tender stem and leaves of quinoa are consumed in salads, side dish, soups or the same way as spinach.

Non-food / Industrial uses: Quinoa is also used in making non-food items as described here under:

- **Animal feed:** Quinoa leaves and tender stems are used as animal feed. Quinoa oil press cake could be an important complementary protein for improving the nutritional quality of animal food stuffs.
- **Green manure:** Quinoa is considered as potential green manure and used for improvement in physical and biological properties as well as fertility of the soil.
- **Cosmetics:** The saponins obtained from quinoa grains may have various potential uses in manufacturing of different cosmetics like detergent, shampoo, soap and body lotion. Traditionally, quinoa saponins is used as natural detergent for washing clothes.
- **Pesticides:** Saponins separated from bitter quinoa grains during processing have pesticidal effects. Therefore, quinoa saponins can be used as pesticide in

integrated pest management program to prevent attack of insects in soil and arial parts of the plants.

- **Pharmaceuticals:** Quinoa is potential source of manufacturing pharmaceuticals.

Medicinal uses

The grain, stems and leaves of quinoa have medicinal value, thus, quinoa is being used as medicinal herb since ancient times by communities of the Altiplano and Andean region for healing purposes. The most frequent uses include treating abscesses, bleeding and dislocations.

(a) Stems and leaves of quinoa are cooked with oil, vinegar and pepper for increasing blood in the body and applied as a poultice to heal sore throats and anginas.

(b) A decoction of the leaves prepared with sugar and cinnamon purifies the stomach, dislodges phlegm, and bile and removes nausea and heart burn.

(c) An infusion of the leaves is used to treat urinary tract infections or as a laxative.

(d) The fresh leaves of quinoa are consumed as a soup, remedy against scurvy and other diseases caused by vitamin deficiency.

(e) Poultices of black quinoa, combined with some other plants, is applied to heal broken bones.

(f) Paste of its fresh fruits made by mixing with alcohol or brandy are used as a remedy for sprains, fractures and dislocations and also as a refrigerant, a diuretic and to prevent colics.

(g) A decoction or poultice made from the fruits is used to apply to wounds and bruises.

(h) Quinoa is good for insomnia, combats dandruff and is a good hair tonic. Likewise, the cooking water from the cooked grain mixed with milk and almond oil is used to wash the ears where there is pain, noise and deafness.

(i) The decoction of quinoa grains, sweetened with honey, is a proven remedy against bronchial disorders, colds, cough and inflammation of the tonsils.

(j) The soup, or warm grain of quinoa is nutritive tonic, increases breast milk and protects against tuberculosis.

17.2 VALUE ADDED PRODUCTS OF QUINOA

Currently several processed / semi-processed value added products of quinoa are available in the markets. Most of them are ready-to-eat products which are usually eaten during breakfast. The details of most common value added quinoa products are given below.

Quinoa flour: Quinoa flour is obtained by grinding quinoa grains from which the saponins have been removed, using pressure and friction, and later airing it to obtain a light powder.

Quinoa flour is used to prepare almost all products of the milling industry due to its high biological value. The main advantage of quinoa as a supplement in the flour industry, is meeting an increase in the international demand for gluten-free products. Due to absence of gluten substance, quinoa flour cannot be used alone for making products, but it is added in some proportions. Quinoa flour can be used in almost all products manufactured by the flour industry and up to 40 percent quinoa flour may be used in making bread, 40 percent in pasta, 60 percent in sponge cakes and 70 percent in biscuits. For bread-making, the suggested ratio is 19 percent quinoa flour and 81 percent wheat flour.

Multi-grain flour: There is a trend of using healthy multi-grain flour by mixing more than one grain. For making a high quality and very effective multi-grain flour, the flour of five grains can be mixed i.e., quinoa flour (50%), oat flour (12%), wheat (10%), barley (15%), Amaranthus (13%).

Fig. 39: *Multi-super grains Flour*

Snack food: Popped quinoa does not appear in the scientific literature. However, popped quinoa is available in North America and used as a snack food or as a garnish, particularly for salads. Quinoa starch is blended with soy flour to produce a deep-fried dry noodles used as a snack food ingredient.

- **Pasta:** Several North American companies currently making and distributing pasta (spaghetti, elbows, pagodas, linguine, shells, and rotelle) products by including quinoa as an ingredient. Northern Quinoa Corporation (Kamsack, SK,) distributes milled quinoa, Gabriele Macaroni Products (Los Angeles, CA,) is one company that produces specialty pasta with quinoa. Quinoa can be added in pasta upto 40% for easy preparation.
- **Noodles:** Noodles are food products being used worldover. Quinoa flour provides an alternative for the noodle and pasta industry. The best ratio for noodles is 21 percent rice flour and 79 percent quinoa flour.
- **Flakes:** Quinoa flaks are available in the markets as a breakfast cereal. To obtain quinoa flakes, grain saponins are first removed using the process for pearled quinoa. The grains are then dried until the moisture content reaches approximately 15–16%. Quinoa flakes are obtained by pressing the grains

between two converging rollers. The size of the flakes depends on the variety and the end use of the product. Quinoa flakes have a wide range of potential uses i.e., in juices combining quinoa with fruits (apple, pine apple and mango); in soups; and in pies, tarts and cakes. For soups and juices, quinoa flakes require less cooking time than the grain, making them easier to use and consume.

- **Backed foods:** Bakery products like different kind of biscuits can be made by mixing the flour of other cereals. Quinoa flour can be used upto 70% in making biscuits. Quinoa does not have gluten forming protein, thus, it cannot be used in making backery products directly without mixing of flours of other grains. Coarse bread called Kispina is produced in South America. In Bread making, quinoa flour can be used up to 30 percent.
- **Cream substitute:** Quinoa starch has a very smooth mouth feel due to small size of starch granules. Various scientists have developed a process to use quinoa starch as the basis for a carbohydrate cream substitute, but this technology has not commercialized till date. The process of isolating pure quinoa starch is also very difficult.
- **Popped grains:** Quinoa grains are also used to make popped grains. Popped quinoa can be used in many ways, including as instant cereals and as a base for energy bars. It has been found that the nutritional quality of quinoa grains (mainly percentage of protein) considerably deteriorated during this process. The process of popping also causes a drop in palmitic, oleic and linoleic acid levels. Grains are popped by putting a handful of conditioned grains (appropriate moisture level) into a clay pot and heating it over a fire fuelled by cow dung. The grain is constantly stirred as it roasts. The roasted grain can be consumed directly or ground into an instant product.
- **Infant and breakfast cereal:** A multi-grain infant cereal and breakfast cereals, including quinoa, are available in the markets of different countries.

Alcoholic beverage: Quinoa is used in preparation of alcoholic beverage with positive impact on digestive system. Such type of preparation is called "Chicha de quinoa" in Andean region.

Above discussed products are new alternatives to increase the use of quinoa in rural and urban families, since they are prepared from quinoa flour instead of wheat flour. Quinoa can be combined with legumes, such as dried broad beans and beans to improve the quality of the diet especially for pre-school and school children through the school breakfast. Currently several processed or semi-processed sub-products are available, although generally at higher prices so that in many cases they become unaffordable for most people.

In short, there is no doubt that the demand for quinoa is increasing, but the total tonnage required is still small. It does allow for a niche market production, but does not lend itself to large-scale production. The organic markets, the non-allergenic markets and non-GMO markets hold the most promise for quinoa grain and quinoa flour. The industrial use is currently limited by small scale production which prohibits quinoa to compete with wheat, rice and barley in the fraction market. There are definite opportunities in the fractions Market for the special attributes of quinoa fractions, especially for the starch component, but they have not yet been realized and exploited by the food industry. The development of a special market for quinoa will probably have to go hand in hand with quinoa processors and industry. The fact that patents for quinoa component products exist, does show that there is some awareness and interest for quinoa fractions within the food industry.

Fig. 40 (a): *Value added food products of quinoa*

Fig. 40 (b): *Value-added food products of quinoa*

Chapter

18

RESEARCH AND BREEDING METHODS OF QUINOA

Quinoa is an underutilized crop that could be a valuable alternative crop to help the world face critical challenges, such as climatic change, food security, human nutrition and over dependance on a few plant species for the world food supply. The breeding objective focus on adapting quinoa to modern agricultture requirements and to commercial processing. The development of widely adoptable varieties of quinoa with high yield and excellent quality of grains is the main priority of quinoa breeding. Therefore, it is desirable to orient quinoa breeding programme toward both food security and market.

Breeding in quiona is a very defficult task due to self-pollinated nature of the crop. The problem is compounded, since the flowers are very small, as a result of which emasculation and hybridization is very tedious and difficult. Inspite of these difficulties, various breeding methods have been practiced in quinoa worldover which mainly includes selection, hybridization and mutation.

18.1 OBJECTIVE OF QUINOA BREEDING

The main objective of breeding in quinoa is the development of a variety with high potential of grain yield accompanied with high protein, low saponin content and highest suitability to Indian conditions. The objective of the quinoa breeding are to increase productivity of grains and leaves, resistance or tolerance to pests and diseases, stability of crop through flexibility in the face of changing weather and improved nutritional quality for food and industries. In quinoa, another relevant object of breeding for genetic improvement is synchrony of grain maturation to enable the adoption of modern agricultural technologies. Despite agronomic potential, quinoa is still an underutilized crop in India and other countries with relatively few active breeding programmes. Breeding efforts to improve quinoa for important agronomic traits are needed to expand quinoa production worldwide. Breeding work may provide the foundation for accelerating the genetic improvement of the crop, with thc objective of enhancing global food security for a growing world population.

18.2 SUITABLE BREEDING METHODS

More than 100 commercial varieties in the Andean region have been developed using diverse breeding methods based on selection and hybridization. Significant progress has been achieved during this process with regards to varieties, ecotypes and germplasm. An easy approach can be the utilization of morphological markers to distinguish the hybrid from the parents. Seeing the emerging potential of the crop and a very limited research work on breeding aspects, there is an urgent need to initiate elaborate breeding programmes in quinoa (conventional as well as biotechnological) for its genetic improvement.

Genetic improvement of quinoa can be achieved by various techniques mentioned below.

A. **Plants Mutation breeding:** Sudden heritable change in the characteristic of plant / organism is called mutation and process of mutation is called mutagenesis. Natural mutation and recombination are the fundamental source of genetic variability used by growers and breeders to adapt these plant species to human requirements and to the different kinds of agricultural systems. Induced mutation is taken place to obtain outstanding genotypes of underutilised quinoa crop. Induced mutagenesis is well adapted and culturally accepted for developing a local cultivar with small genetic changes. In addition, mutagenesis is a simple and cost-effective technology for quinoa breeding. Radiation and chemical-induced mutation techniques have been used successfully in quinoa.

   In mutation breeding, gamma rays and X-rays are most commonly used as physical mutagens, while EMS (Ethyle Methane Sulphonate), EI (Ethyline imine) and sodium azide are the most commonly used chemical mutagens.

B. **Biotechnological approaches:** Modern biotechnology should be designed for the development of excellent varieties of quinoa.

C. **Selection method of crop improvement:** Selection is one of the first breeding methods used for quinoa. It is the simplest and oldest common method of breeding. Natural selection operates in nature without human interferance and nature itself selects the best / fittest plants / organisms. Natural selection has given the cultivated crops and ecotypes in plants.

   Artificial selection is done by farmers or breeders for the purpose of having better crop from a mixed population where individuals differ in characters. Selecting agent exploits the variations existing among the species.

Artificial selection is categorised in following 4 groups:

- **Mass selection:** Mass selection is based on phenotype (external characters) and the harvested seeds are composited without progeny testing. This method cannot bring any new change in the genotype. In this method, most vigorous plants from the mixed population of quinoa crop are selected, threshed together and a mixture of seeds is obtained. The mixture so obtained is sown for raising the new crop of quinoa from which selection is made similarly in the next year. This process of selection is continued till the plants show uniformity in the desired characters and thus constitute a new variety. This method should be applied when breeding for traits with high heritability.
- **Progeny selection:** In this method, superior plants are selected from a heterogenous population on the basis of performance of their progenies under progeny test. The test evaluates the breeding value of a single plant by

the performance of its progeny. This method is not much suitable in case of quinoa as it is good for cross pollinated crops.

- **Pure-line selection:** It is a method of isolating a desirable homozygous individual from the mixed population and multiplying the same without contamination to release a new variety. This method is suitable for quinoa varietal development. Progeny of a single individual obtained by selfing is called a pureline (homogenic strain). Generally, 10 to 11 years are required to produce a variety of quinoa by pure-line selection method.
- **Clonal selection:** A variety that is propagated vegetatively from a single plant is called clone or progeny. Selection of desirable clones from the mixed population of vegetatively propagated crops is known as Clonal selection. Clonal selection is practiced generally with non-flowering or those species which produce seeds poorly or only under special conditions. This method is not suitable for quinoa varietal improvement.

D. **Hybridization:** It is most frequently employed method of plant breeding and suitable for quinoa varietal development. The aim of this method is to quickly recombine favourable traits from several genotypes into a single genotype and produce novel genetic variability. Individual produced as a result of cross between two genetically different parents is known as hybrid. The natural or artificial process that results in the formation of hybrid is known as *hybridization*. When the crosses are made between the plants of same variety, it is called intra-varietal hybridization. When the crosses are made between the plants belonging to two different varieties, it is called inter-varietal / intraspecific hybridization. When the crosses are made between two different species of the same genus, it is called Interspecific / Intragenric hybridization. Transfer of some genes from one species into the genome of the other species is known as introgressive hybridization. The crosses between different species of same genus or different genera of the same family are known as distant hybridization or wide crossing.

Hybridization is an important method of combining characters of different plants. Hybridization does not change genetic contents of plants / organisms, but it produces new combination of genes. Today, it is most common method of crop improvement and vast majority of crop varieties have resulted from hybridization. In hybridization, selection of parents, selfing of parents / artificial pollination, emesculation, bagging, tagging, crossing, harvesting-sorting (F seeds), and raising of F_1 generation are main steps.

E. **Marker assisted selection:** It is an indirect selection process where a trait of interest is selected based on a marker (morphological, biochemical DNA / RNA variation) linked to a trait of interest e.g., productivity, disease resistance, abiotic stress tolerance and quality. This method of breeding based on DNA marker for improving quantitative traits is rare. This method can be useful to select for traits that are difficult or expensive to measure, exhibit low heritability and / or are expressed late in development. At certain points in the breeding process, specimens are examined to ensure that they express the desired trait. The markers used in this breeding method may be morphological, cytological, Biochemical and DNA markers. The term marker is still appropriate to use when directly assaying the gene of interest, because the test of genotype is an indirect test of the trait or phenotype of interest.

Genetically, the first step in this breeding method is to map the gene or quantitative trait locus of interest by using different techniques and then using this information for marker assisted selection. Generally, the markers to be used should be close to gene of interest in order to ensure that only minor fraction of the selected individuals will be recombinants. Generally, not only a single marker but rather two markers are used in order to reduce the chances of an error due to homologous recombination.

F. **Genetic engineering:** Isolation, introduction and expression of foreign DNA in plants /animals is called genetic engineering. It creats transgenic plants, whereby hereditary DNA is augmented by adding DNA from another germplasm sources. It involves introduction of a gene by the techniques of recombinant DNA technology and genetic transformation. It provides an exciting tool to improve specific traits of inbreeds provided a transgene is available for this purpose e.g., suitable version of the cry gene of *Bacillus thuringiensis* has been transfered in maize inbreeds and insect resistant commercial hybrids have been produced using such inbreeds.

G. **Back cross method:** Crossing of a hybrid with one of its parents or an individual genetically similar to its parent, in order to achieve offspring with a genetic identity which is closer to that of the parent, is known as back-crossing. It can be termed as BC1 hybrid and a further cross of the BC1 hybrid to the same parent produces as BC2 hybrid.

H. **Biotechnological approaches:** Recent advances in Biotechnology appears to hold great promise and will probably have a profound effect on varietal development and pest management. Breeders can now introduce genetic variation in quinoa plants by the application of recombinant DNA technology.

I. **Participatory plant breeding:** Participatory Plant breeding is the development of a plant breeding programme in collaboration between breeders and farmers, marketers, processors, consumers and policy makers. This method involves close farmer-researcher collaboraion to bring about plant genetic improvement within a species. Plant breeder can direct their research according to the needs of the specific groups of farmers. This method ensure greater success of adoption of innovation by the farmers.

Following are the main objective of this method of plant breeding:

- Increase production and profitability of crop through development and enhanced adoption of suitable, usually improved varieties.
- Provide a benefit to a specific type of user or to deliberately address the needs of a broader range of users.
- Build farmer skills to enhance farmer selection and seed production efforts.

This method of breeding includes generating genetic variability, selecting within variable population to develop experimental varieties, evaluating experimental varieties, variety release, seed production and popularization and diffusion of information about new variety.

8.3 PRIORITIES OF RESEARCH WORK ON QUINOA

Present research work should be focused on the following needs:

1. Quinoa grains contain a mixture of triterpene glycocides called saponin. Saponin is a bitter tasting compound acting as a general defence against biotic stresses. Although, saponins may be beneficial for plant growth (by deterring herbivory), they must removed before human consumption as they are haemolytic and produce a bitter flavour. Since this process is costly, is often water intensive, and can reduce nutritional value of the seeds, the development of saponin-free lines is a major breeding objective in quinoa. It is found that saponins accumulate in the seed pericarp between 20 and 24 days after anthesis eventually accounting for 4% of the mature seed mass.

   Naturally occuring sweet quinoa strains that contain very low levels of saponins are present within the quinoa germplasm, although the underlying genes regulating the absence of saponins in these lines are unknown. It is urgently needed to identify these genes. Present important need of research is to reduce the content of saponins in the seed hull. In commercial production, removal of saponin from grains require additional cost and hinder direct consumpption of grains.
2. Research is needed to develop different varieties of quinoa widely adaptable to diverse agro-climatic regions with high seed yield and good quality components for food and industrial uses. Such varieties should also be highly suitable for water scarcity and drought prone as well as rainfed areas.
3. Quinoa moth (*Eurysacca melanocampta, Meyrick*) and the ticona complex (*Copitarsia turbata, Feltia sp, Heliothis titicaquensi, Spodoptera sp)* are a major problem in quinoa production. The quinoa varieties resistant to these insects are neither developed nor available in the world. It is, thus, an urgent need to develop insect resistant varieties of quinoa to mitigate major problem of these insects faced globally by the quinoa growers.
4. Downy mildew (*Peronospora farinosa)* is most widespread and destructive disease of quinoa crop and considered potential production constraints particularly when quinoa is introduced in areas outside its traditional growing regions. Quinoa varieties resistant to these most destructive disease are completely lacking. Development of Downy mildew resistant varieties of quinoa are urgently required to tackle major problem of this disease in all quinoa growing areas.
5. Major breeding objective of quinoa development include the development of shorter plants with fewer branches and more compact seed heads, increased heat and biotic stress tolerance and the introgression of the sweet phenotype into commercial varieties. The identification of the likely causative mutation

underlying the sweet phenotype not only provides insights accelerated breeding of sweet commercial varieties using marker-assisted selection. The diversity present in the primary gene pool of quinoa will also help direct fututre breeding strategies.

7. The quality of grains needs improvement with high suitability for transformation into manufactured food products.

Chapter 19

ROLE OF INDIAN MEDICINAL PLANTS MARKETING FEDERATION AND OTHER AGENCIES IN QUINOA DEVELOPMENT IN INDIA

19.1 FIRST ATTEMPTS OF QUINOA INTRODUCTION IN INDIA

Late Smt. Sarvati Devi, A noted social worker, was working on the health improvement of malnourished mothers, pregnant women and children in rural areas of Agra district since 1958. She has been looking forward for nutria-crop as supplement for resource poor people in remote areas. At the suggestions of Americans / Brazilians nutritionists visited rural people of the area in Agra, the focus was gone on quinoa, a crop of Andean Region of Brazil, Peru and Equator of South America. With the help of such visitors of America and Brazil, she arranged seeds of quinoa in 1975 and sown in two different villages of Bichpuri Block of Agra i.e., Chhontana and Nagla Patiram under irrigated conditions. This was the first attempt of quinoa introduction / cultivation in India during *Rabi season* of 1975–76, but, unluckily, she could not get complete success due to lack of technical growing protocol. After this year, nobody has taken further initiation for quinoa development in India. Thus, inspite of nutrient rich super food crop and wide adoptability, quinoa was remain unexplored till 2003.

The second attempt for introduction of quinoa in India was made by Dr. Dharm Singh, the Author of this book, during 2003–04 considering a serious problem of hunger and mal-nutrition in children, lactating mothers and pregnant women in every part of India, long experience on fighting health issues. He arranged 12 demonstrations (2 in each state) on quinoa at farmer fields in Rajasthan, Haryana, Uttar Pradesh, Maharashtra, Gujarat and Delhi under "Pilot Demonstration Programme". Due to such demonstrations, the area under quinoa was increased to about 20 hectares during 2004–05. It was the first time in India when quinoa was successfully grown and established in different states. The full credit of this success was given to the Mother of Poor-Late Smt. Sarvati Devi, who made first attempt for quinoa introduction during 1975–76. She is known as "Mother of Indian Quinoa". Three genotypes (Ancovinto Cancosa, Villarica and Regalona from Chile) were tested using the same growing protocol at all locations.

Inspite of great success under "Pilot Demonstration Programme", nobody taken any further step for successful quinoa development in India, thus, it remains more or less unfamiliar and could not find a place under existing cropping system. The area under quinoa cultivation almost declined after 2004-05. There were following main reasons:

- Lack of authentic knowledge regarding scientific production package, nutritional value and uses of quinoa.
- The lack of publicity and extension systems.
- Lack of marketing.
- Lack of complete un-awareness among farmers and consumers.

19.2 PIONEER AGENCIES AND THEIR ROLE IN INTRODUCTION, COMMERCIALISATION AND EXTENSION OF QUINOA IN INDIA

- **Indian Medicinal Plants Marketing Federation (IMPMF):** Considering a serious problem of hunger and malnutrition in children, lactating mothers and pregnant women in every part of India, long experience on fighting health issues and success of quinoa growing under "Pilot Demonstration Programme", Dr. Dharm Singh, Director General, Indian Medicinal Plants Marketing Federation taken quinoa development issue again in 2007 as he already got success in quinoa production during 2003–04 and 2004–05. He conducted a 5 years study on "Evaluation of productivity and Production potential of quinoa under marginal areas of Indian sub-continent" during 2009–10 to 2013–14 in 75 potential districts of Maharashtra, Gujarat, Uttarakhand, Delhi, Haryana, Uttar Pradesh, Madhya Pradesh, Rajasthan, Himachal Pradesh and Punjab states with a view to assess production potential and promote its cultivation to ensure its availability in the Indian market for direct consumption and to meet out the demand of food industries for manufacturing of value-added products. The findings of this programme revealed that quinoa could be grown successfully everywhere in India due to its climate hardy nature and wide potential of adaptability to harsh environments. The conditions observed favourable for quinoa cultivation include temperature from 15 to 35°C, sandy to sandy loam soils, soil pH from 6.5 to 8.9 and Relative Humidity from 40 to 85% including desert and dry regions. Studies under project also revealed that quinoa can replace traditional low income food grain crops like wheat and paddy and can play a vital role in combating food insecurity, silent hunger and in development of food industries under existing and emferging adverse ecological conditions.

  The five years successful study from 2009–10 to 2013–14 under the project also reveal that two genotypes - Villarica and Sarvati (S-7, a high yielding variety developed under project) shows greater adaption with stable grain yield up to 35 Q ha^{-1}. The highest grain yield of 42 Q ha^{-1} was recorded by Sarvati (S-7) in Haryana and lowest yield of 15 Q ha^{-1} in Rajasthan with Villarica genotype. Extreme hot weather and warm night inhibit fruit setting and slightly frost damage noticed when occur at the time of flowering in areas in all cultivars. Extreme hot weather and warm night inhibit fruit setting in some locations and slightly frost damage was noticed when occur at the time of flowering in hilly states especially in Uttarakhand. Studies under project further revealed that Quinoa can replace traditional low income food grain crops like wheat and paddy and can play a vital role in combating Food insecurity, silent hunger and in development of food industries under existing and emerging adverse ecological conditions.

Federation is playing a pioneer role in breed improvement, varietal development, evolving, refinement and standardization of area-specific scientific production package for different ecological conditions, arranging testing/on-farm demonstration and dissemination of area-specific, cost-effective, adoptable and acceptable production technologies of different medicinal crops including quinoa throughout India.

Long working experience on fighting hunger and malnutrition by quinoa now reveals that India needs several million tones of quinoa per year to meet the requirement of healthy and nutrient rich food for diabetic and cardiac patients, mal nourished children and for the solution for serious problem of diabetes and cardiac problems of present and future generation. This nutrient-rich food crop can also be another lucrative cash crop for millions of farmers as it has high potential for next revolution in agriculture dominated India.

Indian Medicinal Plants Marketing Federation have taken second big step by sponsoring following two national programme:

1. "Super foods for Nutritional Food Security and Doubling Farm-income (SNFDFI)" implemented by "Sarvati Devi Memorial Trust".
2. "Extension Research Systems for Quinoa Technology Transfer" implemented by 3D Industries Pvt. Ltd, New Delhi.

3D Industries Pvt. Ltd.

3D Industries Pvt. Ltd. an ISO 9001-2008 and ISO 14000–22000 certified company, is a fastest growing and most admired Agro-industry in India. It is a leader in innovations, manufacturing and marketing of world's leading organic crop-inputs in India and abroad. This company introducing, popularizing and commercializing quinoa and other traditional super food grains in different states in India under organic farming systems. The mission of 3D Industries is to strive for better world and to emerge as a world leader / global company in development and production of innovative, multi-functional and eco-friendly organic crop-inputs through technological advancement, innovation and customer satisfaction and ultimately to contribute to the welfare of society.

3D Industries Pvt Ltd is pioneer agency in India implementing "Extension Research Systems for Quinoa Technology Transfer" programme sponsored by Indian Medicinal Plants Marketing Federation. Under this programme, 3D Industries Pvt Ltd arranged On-farm demonstrations and Field Days on Quinoa in 8 States to popularize and disseminate sandardized production package of quinoa for diversification in traditional foodgrain producing areas towards very high remunerative quinoa crop. This company also producing quality seeds of quinoa to facilitate commercial quinoa producers in India.

The indiscriminate uses of chemical fertilizer and pesticides resulted harmful effects on the health of human beings, soils, ground water and environment throughout world. Therefore, this company promoting the cultivation of quinoa only under organic farming systems throughout India. For organic cultivation of quinoa, this company is arranging supply of organically certified crop-inputs on 25 percent subsidy to the farmers under the scheme sponsored by the Indian Medicinal Plants Marketing Federation. This sponsored programme will be continued till 2025. This company is also working to explore the Organic Crop Revolution in agriculture sector in India through development and adoption of sustainable, economical, eco-friendly organic farming systems towards a mission of making India completely an Organic country especially in super food production. The organic quinoa revolution is termed as "Gold revolution".

Sarvati Devi Memorial Trust

Sarvati Devi Memorial Trust (A National Voluntary Organization registered on 19th May, 1988 under Indian societies Act of XXI, 1860 in Delhi) was founded in the everlasting memory of late Smt. Sarvati Devi to carry forward her cherished ideals of social reforms, to keep the torch of enlightenment burning and perpetuate her memory. She was a noted social worker, known as "Mother of Indian quinoa and Mother of poorest of the poor, whose noble dreams were to uplift and enlighten the rural life of India. She not only felt the pulse of the countryside but also identified herself with the sufferings and miseries of the rural folks. She hailed from the Taj city of India and she aspired for social reforms without any motive of self-aggrandizement of political mileage.

It is a secular, non-political, non-governmental and non-profit making organization working for socio-economic empowerment of people with a view to improve the quality of life of most vulnerable and disadvantaged sections of society through a multi-pronged and multi-disciplinary convergence of services. The main focus of organization is on health, agriculture, employment, gender equity, education, social security and empowerment of rural communities. Presently, this organization is doing sustained efforts and playing a vital role in upliftment of rural life through its 10 Regional Centres in 10 States (Uttar Pradesh, Rajasthan, Gujarat, Uttarakhand, Haryana, Madhya Pradesh, Himachal Pradesh and Maharashtra).

This organization is implementing National Programme entitled "Superfoods for Nutritional Food Security and Doubling Farm-income (SNFDFI)" sponsored by Indian Medicinal Plants Marketing Federation. This programme was launched on 5th June, 2014 on the occasion of ''JAN KALYAN DIWAS" celebrated on 106th Birth Anniversary of Indian Quinoa Mother - Smt Sarvati Devi. This programme is being implemented in 10 states through its Regional centers on Public Private Partnership (PPP) and Private Private Partnership (PPP) approaches to evaluate productivity and production potential of quinoa and other high value superfood grains under marginal environment for farm income enhancement and livelihood generation. In each state, the multi-location trials are conducted at different locations under different ecological conditions to ensure production and supply of superfood grains including quinoa wih a view to meet out the demand of poor rural people for their direct consumption and also to meet out the demand of food industries for manufacturing of value-added products.

This organization is promoting mainly 4 super foodgrains i.e., quinoa, chia seeds, teff and amaranth through integrated approach which includes:

- Community organization.
- Skill development training to growers of super foodgrains for production, processing and value addition.
- Arranging Training and visit programmes and live demonstrations for producers / farmers of super grains in the project area.
- Formation of Super foodgrains producer's organisations.
- Strengthening of linkages of quinoa, chia seeds, teff and amaranth producers and users with agricultural departments, extension workers, research scientists and leading Food producing and marketing companies to facilitate them.

This organization have arranged various training programmes on production, processing and value addition, Training and Visit programmes, live demonstrations, super foodgrains / quinoa in kitchen programme, seminar / conferences and interface meet of super food grains / quinoa producers with scientists, policy makers, administrators and agriculture officers.

The four years study (2014-15 to 2017-18) under the project "super foods for the Nutritional food security and doubling farm income reveaded that quinoa can be grown successfully in Uttar Pradesh, Rajasthan, Haryana, Himachal Pradesh, Uttarakhand, Gujarat, Madhy Pradesh and Maharashtra where temperature ranges from 15C to 35°C and soils are sandyloam. Study under project also revealed that indian variety i.e., Sarvati (5-7) performed better at all locations. The highest grain yield of 42 q. per ha in Haryana and lowest of 22 g/ha in Rajasthan was recorded with sarvati variety. Early hot weather and wasm night during flowering and fruit setting inhibit grain yield in Rajasthan.

- **Linking of Quinoa producers, traders and consumers in India:** Sarvati Devi Memorial Trust is playing a pivotal role in linking quinoa producers / farmers to Farm Producer's Organizations (FPOs), Quinoa traders / importers, Quinoa marketing Agencies and Quinoa development Associations at national-level. The purpose of this linking is to extent technical guidance to the farmers and provide market support to them and brought quinoa producers and traders of India on one platform. With this object, "Sarvati Devi Memorial Trust" constituted a National-level Association named "Federal Union of Quinoa growers and Traders of India" on 4th June, 2016, a grand occasion of 108th Birth Anniversary of Late Sarvati Devi, Mother of Indian quinoa. This National level Association is linking quinoa producers to food companies to avoid middleman and ensure attractive market prices. This Union is rendering all need-based services directly to quinoa growers and consumers. It has registered 322 Quinoa producers, 87 Quinoa traders, 23 Quinoa processing machine manufactures and 16 Quinoa-based Food industries as on 30th March, 2018. Anyone having concern with quinoa can get membership of this Association by filling an application form and prescribed Annual fee.

For getting any kind of support from FUQGTI (A national level platform), it is primary necessity to obtain membership by filling application form in prescribed formate. Other necessary information is given below:

- There are 11 members in National Governing Council of National level quinoa federation i.e., Chairman, Vice-Chairman, Secretary General, Secretary, Director-Finance, Director-Administration and other 5 as Executive Members.
- Executive Members in National Advisory Council can be quinoa researchers, quinoa growers, progressive farmers, quinoa traders, exporters or any of agriculture background.
- Membership fee will be as under:

For farmers:

- Admission fee : Rs. 25/-
- Annual subscription fee : Rs. 100/-

For Corporate sector and traders:

- Admission fee : Rs. 500/-
- Annual subscription fee : Rs. 2500/-
- There shall be no limit of members of General Body of this federaion.
- All the members of National Governing Council of this federation will be elected / Nominated from the General Body of Sarvati Devi Memorial Trust and members of this federation.

GUIDELINES FOR PROCUREMENT OF QUINOA / OTHER SUPER GRAINS

- Sarvati Devi Memorial Trust will implement "Production programme of super foodgrains" in India with more emphasis on 8 states i.e., Maharashtra, Madhya Pradesh, Gujarat, Uttar Pradessh, Delhi, Haryana, Uttarakhand and Rajasthan states.
- Sarvati Devi Memorial Trust will not procure quinoa directly from farmers, but it will act as facilitator and provide all kind of support to the growers under programe. Quinoa or other super grains will be procured from farmers by Federal Union of Quinoa growers and traders of India on behalf of 3D Industries Pvt. Ltd under full guidance and rules and Regulations approved by the Sarvati Devi Memorial Trust, so that farmer's interest could be secured.
- Sarvati Devi Memorial Trust has approved and fixed Rs. 100/- as Minimum Support Price (MSP) of organically grown quinoa for procurement in different states. It may be revised year to year as per need.
- About Rs. 10 to 15 per Kg may be given additional to the quinoa growers as bonus for A - Grade quinoa grains. The decision regarding bonus will be taken by the competent authority of Sarvati Devi Memorial Trust after drawing a sample and quality testing of produce of each farmer.
- It will be the responsibility of farmers for cleaning and grading of quinoa grains properly before procurement / sampling.

- Representative of Sarvati Devi Memorial Trust will draw sample of quinoa grains about 10 days before procurement to decide about MSP / bonus to be paid.
- Payment of farmers against their produce will be given within 30 days from the date of procurement / lifting of quinoa produce.
- Quinoa will be procured only when farmers will fulfil the following criteria:
  - When quinoa is grown organically.
  - When inputs used by the farmers are organically certified and approved by Central Insecticide Board or supplied by Sarvati Devi memorial Trust.
  - When quinoa producers / farmers are member of Federal Union of Quinoa Growers and Traders of India (FUQGTI).
  - When quinoa producer / farmer follows production package / protocol provided by Sarvati Devi Memorial Trust.
  - When producer / farmer have cleaned and graded his produce well before procurement.
  - When producer / farmer will pack his produce in new gunny bags having plastic layer inside in presence of representatives of Sarvati Devi Memorial Trust.
  - When Farmers will complete their all post-harvest operation (threshing, cleaning and grading) and arrange their supply qualiy testing.
  - When quinoa producer / farmer executed an agreement with 3D industries Pvt. Ltd. or Federal union of quinoa growers and Traders of India.

Benefits to members of FUQGTI

- Member farmers shall be benefitted through linking them with various Govt. Schemes.
- Member farmers shall be benefitted through quinoa based activities i.e., organizing skill development trainings, vocational trainings and awareness camps for income and crop productivity enhancement.
- Member farmers shall be benefitted through organizing health checkup camps free of charges.
- Member farmers shall be benefitted through door-step delivery / availability of organically certified quality inputs.
- Member farmers shall be benefitted through organizing exposure visits in State / India / Abroad.

Quinoa Producer's Organisations / Associations

- Sarvati Devi Memorial Trust is playing major role in formation of quinoa producer's organisations / quinoa development associations and other agencies engased in quinoa development in each state with a view to link quinoa growers, users, food companies, traders / exporters.

Chapter
20

BRIEF ABOUT SUPER FOOD GRAINS OTHER THAN QUINOA

Super foods don't have their own food group and dieticians, nutritionists and scientists do not use this term as it is not scientific. According to Cancer Research, UK, the term "Superfood" is just a marketing term used to describe foods with supposed health benefits. According to American Heart Association, there are no set criteria for determining what is and what not superfood is. Super foods have extra-large doses of soluble fibre, vitamins, minerals and phyto-chemicals that can help us ward off diseases and live a longer life.

The Oxford Dictionary states that super food is nutrient rich food considered to be especially beneficial for heath and well-beings. Super food grains are a special category of foods found in nature. By definition, they are calorie sparse and nutrient dense, meaning they pack a lot of punch for their weight as far as goodness goes. They are superior source of anti-oxidants and essential nutrients we need. When we levelled foods as super and healthy, people think they can eat them in unlimited quantities. But, you do have to be cautious of the amount you eat, because you can gain weight from eating too much healthy food.

Whole wheat and oats are a mainstay in most of our whole grain recipes, but there are other lesser known grains that are not only delicious, but packed with nutrients. Our body need protein to create new cells, repair old ones and general development throughout our lives. Our body can more easily break down vegetable proteins into their constituent amino acids, therefore, some nutrient rich foods can be treated like super food grains which are an important quality food for everyone, not just vegetarians but also for healthy food fans. Super foods are highly suitable for mid-day meal programme.

The most important super grains are described hereunder in brief.

*1. Amaranth (**Amaranthus species**)*

Amraranth is a common name for more than 60 different species of Amaranthus. Three species i.e., *Amaranthus cruentus*, *Amaranthus hypochondriacus* and *Amaranthus caudatus* are commonly grown for their edible seeds. Amaranth is a very rich source of complete protein which is highly bioavailable and more digestible than other food grains and has been compared to the digestibility of milk protein. Amaranth has a good amount of lysine which helps the body absorb calcium, build muscle, and produce energy. Amaranth has been cultivated as grain for about 8000 years. It is not a cereal but classified as pseudo-cereal, which means that belongs to a different plant species grown for its edible starchy seeds. Amaranth is considered a weed by much of the world, but it has been grown as a grain-like seed and a vegetable for thousands of years. It has about the same yield as many other commonly used grains like rice and a similar nutritional value to quinoa, which makes it well worth looking into. Its seeds are also used to extract oil for commercial use. Amaranth is grown mainly in Mexico, China, Thailand, Russia, Nigeria, India and parts of South America.

Specialities of Amaranth

- It is rich source of protein containing about 13 – 14% with highest digestibility.
- Its protein is referred as complete as it contains lysine, an amino acid missing in most of common food grains, which helps in absorption of Ca, build muscle and produce energy.
- The significant level of carotenoids and vitamin A found in amaranth leaves is a major boost for eye health, as these antioxidants can prevent macular degeneration and slow/stop the development of cataracts. By lowering oxidative stress in the ocular system, amaranth can help to keep your vision healthy and strong for years to come.
- It is cholesterol lowering whole grain. Oil and phytosterols reduces total cholesterol, tryglycerides and LDL (bad) cholesterol.
- It is naturally gluten free.
- Peptides can ease pain, reduce inflammation and prevent cancer. Phyto-nutrients and fibre lower blood pressure and making it good food for heart.
- It is rich in minerals like Ca, Mg and Copper.
- It is rich in vitamins including A, C, E, K, B_5, folate, niacin and riboflavin.
- Folate is often overlooked in terms of essential minerals, but it is particularly important for pregnant mothers. Folate deficiency can result in neural tube defects in newborns, so add some amaranth grains or leaves to your diet and protect your newest little addition to the family.
- Amaranth helps to prevent premature graying, mainly due to the minerals this seed has in abundance.
- Amaranth contains a rare amino acid called lysine that the body cannot naturally produce. This increases your calcium uptake efficiency and helps to keep the hair on your head, strengthening follicles and preventing male pattern baldness. Juice of the leaves, if allied to hair after shampooing, strengthen hair and prevent hair loss.
- Amaranth have anti-inflammatory properties due to peptides and oils which can ease pain and reduce inflammation. This is especially important for chronic conditions where inflammation erodes your health, such as diabetes, heart disease, and stroke.
- Peptides in amaranth protect against inflammation which may also help prevent cancer. The antioxidants in this grain may even help protect cells from other damage that can lead to cancer.

- The fibre and phytonutrients in amaranth lower blood pressure according to some recent studies. This seed tackles cholesterol, inflammation and blood pressure, making it an all-around good food for heart health.
- Amaranth contains numerous flavonoids, including rutin, which has been directly connected to eliminating varicose veins by strengthening capillary walls. This is also aided by amaranth's high concentration of vitamin C, an integral component in the production of collagen, which also helps repair and strengthen blood vessel walls.

Fig. 41: *Grains of Amaranth (Amaranthus species)*

*2. Chia (**Salvia hispanica**)*

Chia seeds rank among the top plant based sources of protein. It is drought hardy crop originated in central and southern Mexico and Guatemala, and grown in Mexico, Guatemala, Bolivia, Argentina, Australia, Nicaragua and Equador. It is used in making breads, energy bars, and breakfast cereals. Chia seeds are generally used to replace egg content in cakes. It assists digestion of other foods. Chia seeds contain large amounts of fibre and omega-3 fatty acids, plenty of high quality protein, and several essential minerals and antioxidants. Chia seeds are small, flat and oval-shaped, with a shiny and smooth texture. The colour can range from white to brown or black .

This super seed is great to consume for those trying to put on lean muscle, burn fat, and balance blood sugar levels. It also improves stamina and endurance, making chia a great source of nutrition if you are looking to get in shape. One of the characteristics that make chia so unique is that it can absorb water up to 12 times of its own weight. Because of this, chia seeds can prolong hydration and improve nutrient absorption of electrolytes. This also slows digestion and keeps you feeling fuller longer reducing sugar cravings. Chia seeds are so rich in antioxidants and they can be kept in storage for up to four years without deterioration.

Fig. 42: *Chia seeds (Salvia hispanica)*

Specialities of chia seeds

Chia seeds contain 486 calories per 100 grams, water 6% , carbohydrates 46%, fat 34% and protein 19%. Carbohydrade contains morethan 80 percent fibre which is insolube. About 75 percent of the fats in chia seeds consist of the omega-3 fatty acid alpha linolenic acid (ALA), while about 20 percent consist of omega-6 fatty acids.

- Chia seeds are rich in manganese, phosphorus, copper, selenium, iron, magnesium and calcium nutrients.
- Chia seeds contain chlorogenic acid (anti-oxidant), Caffeic acid (fight inflammation), quercetin (reduce the risk of heart disease, osteoporosis and certain forms of cancer, and Kaempferol (reduce risk of cancer and chronic diseases).
- When chia seeds are placed in water or other liquids, the fibres absorb up to 10-12 times their own weight, and the seeds turn into a gel-like mass. It helps to quench the thirst in summer when added in a glass of water.
- Chia seeds are one of nature's richest antioxidants which speed up the skin's repair systems and prevent further damage. Taking chia seeds can prevent premature skin aging due to inflammation free radical damage.

- It is super-high in fibre, which is essential for human body's ability to balance insulin levels and can be a natural blood sugar balance.
- When chia seeds consumed, it create a gelatin-like substance in the stomach due to soluble fibre and curbs hunger and suppresses appetite.
- Chia reduces weight.
- Chia seeds capability to reverse inflammation, regulate cholesterol and lower blood pressure make it extremely beneficial to consume for heart health.
- Chia seeds are high in healthy fats like Omega-3 and linoleic fatty acids which helps the body to absorb fat-soluble vitamins A, D, E and K.
- It can help prevent metabolic disorders like dyslipidemia (excessive fat in the blood) and insulin resistance which are two factors in the development of diabetes.
- Chia seeds enhanced exercise performance for workouts that lasted 90 minutes the same way a sugar-laden sports drink would, but without all the sugar.
- It helps in boosting metabolism and burn belly fat that effects the metabolism of the body and is a component of obesity.
- Chia seeds are rich in alpha lineic acid / alpha linolenic acid (ALA) which is an Omega-3 fatty acid. As per recent studies, ALA limited the growth of cancer cells in both breast and cervical cancers.
- Chia seeds are rich in calcium which is the building block of teeth and necessary for tooth health.
- Chia seeds contain almost five times the amount of calcium as milk and also contain boron. Both are critical mineral for bone health and beneficial in skeleton development of baby during pregnancy.
- Chia slows down the process of converting sugars and carbohydrates from the seeds into energy. This slow-burning process, in combination with the high protein content, leads to a sustained energy boost.

*3. Teff (**Eragrostis tef**):*

Teff (*Eragrotis tef*), is native to Ethiopia in Africa which is the largest teff producing country in the world. It is grown for its grain used in making the staple, in more than 50 percent area of cereal crops in Ethiopia. It is also grown on small scale for food in Yemen, Kenya (near Marsabit), Malawi and India. Inspite of increasing market for its grain, teff is mainly a forage crop in the US, South Africa and Australia. It may become the new "super grain" of choice in Europe and North America, overtaking the likes of quinoa and spelt. It is more expensive than quinoa.

Teff is adapted to environments ranging from drought stress to waterlogged soil conditions. Teff is daylight-sensitive and flowers best with 12 hours of daylight. Maximum teff production occurs at altitudes of 1,800 to 2,100 m, 450 to 550 mm rainfall and 10 to 27° C temperature. Teff is very intolerant of cold temperatures and frost.

Fig. 43: *Teff (Eragrostis tef) grains*

Teff is a much smaller grain compared to rice and wheat, and it typically cooks faster than most grains. This makes teff an energy-efficient grain, making it popular among environmentalists and health foodies alike. It is not only rich in magnesium, iron, fibre, Vitamin B6, calcium, and protein, but it is also the only ancient seed to contain vitamin-C which can help to protect our immune system against deficiencies, lower risk of heart disease and can even prevent premature ageing. Teff flour contains five times as much fibre as compared to wheat flour.

Teff is a gluten-free grain so it can be a great alternative for those living with celiac disease, having gluten intolerance or choosing a gluten-free life-style. It is used in gluten-free baking bread, in soup as thickener, in stews, gravies, puddings and stir-fry dishes.

Specialities of teff

- Teff contains approximately 20 to 40 percent resistant starches and has a relatively low glycemic index (GI) that can help diabetics, better regulate their sugar levels.
- Teff is very low in saturated fat.
- It is a great source of gluten-free protein. Gluten containing protein is responsible for allergic and digestive reactions in the human body.
- Teff is not a complete source of protein, but, it is a great source of 8 essential amino acids present in higher amount as compared to other grains. Lysine content is higher than those of wheat or barley.
- Carbohydrates of teff are slowly broken down into glucose for a steady assimilation into cells. This means the carbohydrates of teff are slowly absorbed by the body, which prevents large fluctuations of the blood sugar level.
- Regular intake of teff grains remove toxic wastes from the colon, which may be helpful for reducing accumulation of toxic compounds in the body.

- Teff provides about 30 percent of daily value of copper, which supports healthy weight and energy production. Copper also increases energy levels, being involved in the synthesis of ATP (energy molecule of the body that is produced in mitochondria). Copper also helps liberate iron from protein.
- Teff grains are loaded with manganese, which helps with producing digestive enzymes responsible for a process called gluconeogenesis. Gluconeogenesis supports energy, breaking down protein into amino acids, and then into sugar for fuel. This process supports balanced blood sugars and stable energy.
- Soluble fibre of teff eliminate more cholesterol particles and fats before they reabsorbed from intestines. Insoluble fibre is not absorbed, and thus passes through digestive tract, taking with toxins, metabolic waste, damaged fats and cholesterol particles. It promotes digestive regularity as well as nutrient absorption, making it vital component of a healthy diet.
- Teff is full of iron which is easily absorbed by the body. It helps in controlling blood sugar levels.
- Teff grain has been known as an anti-inflammatory and menstrually soothing agent. Thus, it prevent heavy menstrual flow, severe cramping or other physical manifestations of the monthly event.
- Teff is a boon to keep away anemia especially in women who are prone to anemic conditions.
- Teff grais are good source of calcium and magnesium which are required for healthy and strong bones. Teff is considered good for people suffering from osteoporosis.
- Teff is good source of dietary copper and zinc, which are essential in preventing premature hair loss.
- Teff grain contains phytosterols which are promising in research of cancer. These phytosterols prevent secretion of estrogen hormone in body which promotes breast cancer.
- Teff grains are an anti-inflammatory and menstrually soothing agent, so helpful to those who suffer from heavy menstrual flow, severe cramping, or other physical manifestations of the monthly event.

*4. Spelt (**Triticum aestivum var. spelta**)*

Spelt is a sub-species of common wheat and it has been considered an undeveloped crop. It is often erroneously called "speltz". Sometimes, another sub-species of wheat that includes durum wheat, is incorrectly called spelt. Switzerland and Germany are considered its origin. Spelt adapts well to cold / temperate climates and thrives in various soil types like sandy, wet, poor and non-draining. It is less demanding for nutrients and more competitive against weeds than wheat, which make it an interesting cereal crop for mountain areas. Spelt has been grown in Europe for about 300 years especially in Switzerland, Germany, Belgium and Spain. It was introduced to the United States in the 1890s and it is mostly grown in Ohio.

Germans use the grain as animal feed, in bread and for brewing. Ground spelt is used as an alternative to oats and barley in animal feed. Its nutritional value is similar to that of oats. In addition, spelt hulls have almost as much nutritional value as the grain. De-hulled spelt can be consumed by humans and is a popular fibre source in Europe. In the past decade, the demand for spelt has increased substantially in the U.S. food market and gained widespread popularity as a wheat substitute for making artisan breads, pasta and cereals. Spelt can also be used in flour and baked goods to replace soft red winter wheat. Spelt contains more protein, fats and crude fibre than wheat and also has large amounts of Vitamin B17 (anti-carcinoma). The protein content of the US based Champ variety of spelt is about 11.7%, compared to 12 to 13 percent for oats. American food manufacturers have begun to use spelt to meet the nation's increasing demand for pasta and high fibre cereals.

Fig. 44: *Grains of Spelt (Triticum aestivum var. spelta)*

Spelt is packed with nutrients that many other cereals do not contain. It is high in carbohydrates and very high levels of protein (21% of our daily recommended intake) and dietary fibre (30% in a single serving) are very impressive. It also contains special carbohydrates which play a decisive role in blood clotting and stimulate the body's immune system so as to increase its resistance to infection. Spelt contains about 70% total carbohydrates, including different text in both book. Fat content is low in spelt. Spelt's nutritional vital substances are found in the grain's inner kernel with high water solubility, thus, absorbed quickly into the body, aiding digestion. The body cells are then nourished, strengthened and prepared for their optimal performance, while the body is flooded with vitamins and other nutritional substances.

Specialities of Spelt

- Spelt contains a moderate amount of gluten and is therefore suitable for baking, but this component makes it unsuitable for people with gluten related disorders, such as celiac disease.
- Spelt is rich in protein which contain all nine essential amino acids needed by the human body. It varies from 13.1 – 14.28 percent, higher than soft wheat (10.5%) and spring wheat (9.1%), but similar to durum wheat (13.8%).
- Spelt also contains significant levels of iron, copper, manganese, magnesium, phosphorous, potassium, zinc, selenium, niacin, thiamin, vitamin B6 and folic acid. When compared to wheat flour, spelt has higher contents of copper, iron, zinc, magnesium and phosphorus.

- In comparison to hard red winter wheat, spelt has a more soluble protein matrix characterized by a higher gliadin : glutenin ratio.
- Spelt has one of the highest dietary fibre contents over wheat. Dietary fibre helps to bulk up our stool and move food through the digestive tract, speeding up the absorption of nutrients and helping to reduce conditions like constipation, bloating, cramping, excess gas, diarrhea and more serious gastro-intestinal issues like ulcers.
- Dietary fibre has a second role in the body, which is to help lower the levels of dangerous cholesterol. Dietary fibre can interact with the cholesterol uptake processes and inhibit the absorption of cholesterol from food. In fact, it can specifically target LDL cholesterol and eliminate it from the body to regulate a healthy balance of fatty acids in the body.
- The high levels of iron and copper both significantly boost circulation. These nutrients are essential for the production of Red Blood Cells. When RBC production is up, then, blood flow get increased throughout the body. It means additional oxygenation to organs and tissues which increases healing, boosted energy levels and a more functional metabolism. Increased circulation can even boost hair growth, while the iron content helps the body avoid becoming anemic.
- Niacin, a part of the vitamin B complex, is one of the essential water-soluble vitamins that can be found in significant quantities in spelt. Niacin plays a key role in the adrenal glands in the body, particularly in the production of sex hormones.
- The endocrine system is a sensitive and hugely important aspect of our health and general functioning, so maintaining healthy niacin levels by adding spelt to your diet is a wise choice. It is important for maintaining a healthy cardiovascular system and metabolism-especially balancing blood cholesterol levels. Niacin helps with brain function, healthy skin formation, and preventing or treating diabetes. Niacin can help to lower the risk of developing neurodegenerative diseases, including Alzheimer's disease, memory loss, eye disorders like cataracts, acne and skin flair-ups, osteoarthritis, circulation problems, migraine headaches, dizziness, and learning disorders like ADHD.
- Thiamin is known to significantly boost our immune system and stimulate the defensive mechanisms of the body, also easing psychological stress and anxiety. Spelt is one of the few cereals that can boost a significant level of vitamin–B.
- Spelt has an impressive range of essential minerals i.e., zinc, magnesium, copper, phosphorous, and selenium. These minerals contribute to the development of bone tissue and by boosting the levels of these minerals in our body, osteoporosis and other age-related conditions that weaken or degrade the bones in our body can be prevented.
- The high levels of phosphorous, combined with the impressive content of protein in spelt makes this cereal very important for the development and growth of new tissues, blood vessels, muscles, bones and organs. Phosphorous is one of the essential elements behind creating new proteins and making DNA to build new cells, while the proteins you get directly from spelt can be

broken down into their composite amino acids and then re-formed to be used in the body for nearly every important bodily process.

- Foods high in phosphorus, like spelt flour, also maintain the body at the proper pH level and help with energy extraction.
- Although simple sugars and carbohydrates are known to be dangerous for diabetics, the high fibre content in spelt can counteract those effects and help to regulate the release and breakdown of those simple sugars into glucose. By regulating the release of insulin and glucose in the body, it helps to manage diabetes for those who already suffer from the condition, or prevent more people from developing it.
- The vitamins and minerals in spelt flour help to boost the immune system and reduce inflammation.
- Iron helps to bring enough oxygen to damaged areas of the body, including damaged tissues, organs and cells that are prone to infection or disease development. Iron helps to metabolize proteins and plays a role in the production of hemoglobin and red blood cells, serving as a natural treatment for anemia.
- Carbohydrate rich foods like spelt flour, doesn't contribute any carbs to our diets. Due to its structure and our inability to absorb it, fibre passes through our digestive system unabsorbed by digestive enzymes within the stomach — taking with it toxins, waste, fat and cholesterol particles and removing them from the gut. In the process, it helps to improve our digestion and heart health, makes good feel and supports detoxification. This is because prebiotic fibre helps to improve immune function and maintains better colon and intestinal health, while also clearing away harmful waste from the digestive organs. Fibre targets LDL (bad) cholesterol and eliminates it from the body in order to regulate the balance of fatty acids.
- Spelt flour helps to regulate the amount of glucose and insulin that's released in the body; this is due to its high fibre content as well. A high-fibre diet may help to prevent type 2 diabetes, lower insulin and blood sugar levels and improve cholesterol and triglyceride (fats in the blood) levels in people with diabetes.
- When spelt due to various minerals, including manganese, calcium, zinc and copper, it minimizes bone loss, especially in older women who are more susceptible to bone fractures and weak bones. By improving bone mass, manganese helps to prevent and treat osteoporosis, which is when small holes or weakened areas are formed in the bone that can lead to fractures and pain.

*5. Finger millet / Ragi (**Eleusine coracana L**)*

Finger millet, also known as African millet, Ragi, Rajgira, Nanchani and Mandua, is an important food crop next to rice, wheat and maize. It is an important staple food in parts of eastern and Central Africa and India. It is believed that finger millet is originated in India, from where it spread to Arabia and Africa, nearly 3,000 years ago. Considering its predominant cultivation in southern India, this region is accepted to be the primary centre of its origin. It is valued as staple food consumed in South India especially Karnataka, Rajasthan, Goa, Andhra Pradesh, Odisha, Tamil Nadu, Maharashtra, Garhwal and Kumaon (Uttarakhand) and Dang district (Gujarat).

It is a very hardy crop predominately grown as a dry land areas in the peninsular Indian states of Karnataka, Andhra Pradesh and Tamil Nadu with rainfall up to 100 cm. It is grown from about 500 meters above sea level up to about 2400 meters above sea level. Ragi flour is made into flat breads, including thin, leavened dosa and thicker, unleavened roti. Ragi grain is malted and the grains are ground. This ground flour is consumed mixed with milk, boiled water or yoghurt. In the Malnad region of Karnataka, the whole ragi grain is soaked and the milk is extracted to make a dessert known as «Keelsa». Ragi is also used to make roti, idli, dosa and conjee. Ragi is traditionally fed to women as a medicine after child birth.

Ragi is a widely cultivated crop of the tropical and subtropical regions of the world, being grown in Africa, Madagascar, Sri Lanka, Malaysia, China and Japan besides large area in India. In India, it is main cereal crop which is extensively grown in Karnataka, Tamil Nadu, Andhra pradesh, Odisha, Bihar, Gujrat and Maharashtra and in the hilly regions of Uttar Pradesh and Himachal Pradesh. Karnataka is the leading producer of ragi in the country where it is extensively grown in Kolar, Bangalore, Chitradurga, Tumkur, Hassan and Mysore districts which together provide 85% of the total area and 80 percent of its output in the state. Tamil Nadu occupies second place in respect of area (7.52%) and production (14.60%) of ragi in India. The chief producers are Coimbatore, Dharmapuri, Ramanathapuram, Salem, North and South Arcot, Nilgiris, Chingleput and Madurai districts. Other important producers of ragi in India include Odisha, Maharashtra (Ratnagiri, Kolhapur, Thane and Raigad districts), Uttara Khand, AndhraPradesh (Chittoor, Srikakulam, Anantapur, Vishakhapatnam, Prakasam, Mahbubnagar and Cuddapah districts), and Bihar.

Fig. 45: *Grains of Finger millets (Ragi) (Eleusine coracana L)*

Ragi is a crop of tropical and subtropical climate, short-day plant which requires a fairly high temperature, ranging from 20 to 40°C. The optimum temperature 32°C during day time and 25°C during night time seems to be more favourable for the development and growth of ragi crop. The crop possesses good drought tolerance but is highly sensitive to frost and yield reduces at temperatures below 20°C. Bright sunshine with low temperature during ripening period of the crop helps in the development of the grains. Heat tolerance of finger millet is high. During pollination, strong wind may induce sterility and increase the number of abortive endosperms.

It can be grown under rainfed as well as irrigated conditions. It grows best in moist climate. It is generally considered as a drought tolerant crop. It does not tolerate heavy rainfall and requires a dry spell at the time of grain ripening. Heavy down pour at anthesis and flowering leads to sterility. It is also raised as summer crop, and as *Rabi* crop in south India, but mainly as Kharif crop in north India.

Finger millet can grow on various soils, can tolerate soil salinity, alkalinity (pH 8.2) and moderately acidic soils (pH 5) up to a certain extent.

Table 16: *Nutritional value of finger millet per 100g*

S.No.	Nutrient	Amount
1	Protein	7.6g
2	Fat	1.5g
3	Carbohydrate	88g
4	Calcium	370mg
5	Vitamin A	48mg
6	Thiamine (B1)	33mg
7	Riboflavin (B2)	0.11mg
8	Niacin (B3)	1.2mg
9	Fibre	3g

Specialities of Finger Millet

- Finger millet helps in weight loss in a unique way and holds a significant place in diet programs for weight loss.
- Ragi is an excellent plant source of natural Iron. Its consumption helps in Anemia. Vitamin-C increases iron absorption in the body.
- Ragi is a low fat cereal in which most of the fats are in the unsaturated form. This makes it the best choice in grains for people trying to lose weight.
- Ragi contains an amino acid, tryptophan which reduces appetite.
- Ragi is loaded with calcium and vitamin D that makes it a vital component for increasing bone strength.
- Ragi works as a super calcium supplier for growing children, aging people, pregnant ladies, working women etc. It helps development of bones in children, maintains bone health and prevents bone erosion in adults. It also works as

an effective food to avert osteoporosis. Consumption of ragi helps to prevent arthritis, could reduce risk of fractures and to recover from fractures faster.

- The high level of dietary fibre in this cereal aids in proper digestion, normal bowel movement and prevents constipation. The insoluble fibres of ragi assists movement of food through the intestines and retain water thereby easing the passage of waste. Thus, it protects our digestive system.
- The bran of finger millet causes indigestion, hence it needs to be soaked, powdered and the bran/husk is removed, then it is dried and consumed to prevent indigestion problems.
- The abundance of antioxidants, mainly tryptophan and amino acids in ragi help in fighting damage causing free radicals and reducing oxidative stress. One amazing benefit of ragi is that it helps in relaxing the body naturally.
- Ragi is very beneficial for diabetic patients. The low glycemic index in ragi slows down the digestion process, keeps the blood sugar level at a constant ratio, increase insulin sensitivity for people suffering with diabetes and also helps to control the sugar levels for non-diabetics especially type-2 diabetes. Thus, regular consumption of ragi can decrease fasting glucose by 32 percent and can eliminate insulin resistance by 43 percent.
- It is found to be beneficial in conditions of anxiety, depression, hypertension, migraine and insomnia. Tryptophan in millet raises the serotonin level in the body which helps with stress reduction.
- Consumption of millet regularly helps to decrease triglyceride levels in the body. It thins the blood to prevent blood platelet clumping, thereby reducing the risk of sun-stroke and coronary artery's disorder.
- Vitamin-B in millet helps to break down carbohydrates and fat more efficiently. It reduces homo cysteine level in the blood to prevent cholesterol from bonding and forming deposits. Niacin prevents cholesterol from getting into the blood stream and raises high density lipoprotein (HDL) in the blood. This protects the blood vessels from atherosclerosis and hemorrhage.
- Ragi has been proved to work wonders in bringing down blood cholesterol levels and is one of the best home remedies for heart health. It controls the cholesterol level by reducing plaque formation and preventing blockage of blood vessels thereby cutting down the risk of stroke and other heart diseases. The key amino acids Lysine and Methionine present in the cereal helps in lowering cholesterol by extracting and cutting out excess fat from the liver.
- Magnesium in ragi relaxes the muscles that line the inside of the arterial wall, which helps to reduce blood pressure.
- The high amount of antioxidants present in this millet fights free radicals present in the body which slows down the ageing process. Lipoic, an antioxidant found in ragi, produces valuable anti-inflammatory results by increasing cell metabolism and circulation. This anti-inflammatory property increases blood circulation in the body to reduce acne and fine lines on the skin. Consuming finger millet helps to retain the natural moisture content in the skin and makes it healthier and fresher.

- Finger Millet is rich in Vitamin-E, which penetrates through the layers of the skin, assisting the body with the natural wound healing process. It even lubricates the skin and forms a protective layer that serves to keep the free radicals away from the wound.
- Scar tissue has a different cell structure than that of healthy skin, making it rigid. Alium, an antioxidant found in this millet, boosts the blood flow to scar tissue. It blends in the development of new skin, minimizing the scars. It increases the speed of the skin repair system, firms the skin and prevents further damage.
- Antioxidants in this millets fight stress and neutralize free radicals in the body. This helps to reverse the signs of ageing on the skin. It maintains proper health of the skin by rejuvenating skin cells. Ubiquinone in this millet is also used in beauty products to reduce wrinkles on the face.
- Ragi not only helps in maintaining proper health and keeping diseases at bay, but also works wonders for maintaining youthfulness and vitality of skin by blocking premature ageing. Ragi can be used to be the most effective natural anti-aging drink.
- The vital amino acids methionine and lysine present in this cereal help in creation and maintenance of collagen. Collagen is a substance that makes skin less prone to wrinkles and sagging.
- Pregnant and lactating women are advised to consume ragi as it helps to increase the production of breast milk in the body. This will enable the mother to feed the child for a longer period of time. Ragi proves to be a great cereal for pregnant and lactating women due to its high calcium and iron content.
- It is rich in folic acid which helps in brain development of baby and calcium in ragi will helps in bone development of child. Ragi is also rich in dietary fibre and hence bowel friendly.
- Ragi is a great source of nutrition for new mothers as it helps in improving the hemoglobin levels. Lactating mothers are suggested to include Ragi in their daily diet in order to improve milk production.
- It also enriches the milk with all the essential minerals for the body like iron, calcium and vital amino acids that is extremely important for the child as well as the mother.
- Due to its high calcium content and easy digestibility, its grains are malted and used for weaning babies. Ragi helps to fulfill the necessary mineral requirement of growing children and therefore helps in proper growth and development.
- Finger Millets are rich in protein, a nutrient highly recommended for people suffering from hair loss. Healthy hair requires sufficient intake of protein, as most of the hair itself is made of protein. The protein found within hair cells is known as keratin, which is directly responsible for the composition of each strand of hair. Lack of protein can lead to severe hair loss. Thus, an adequate intake of this millet strengthens hair, making it stronger and less prone to breakage.

- Magnesium in millets reduces inflammation in the scalp to treat scalp conditions like eczema, psoriasis and dandruff. It ameliorate metabolic syndrome, a health condition which causes premature balding. Millets promote blood circulation in the scalp and promote hair growth.
- Stress increases cortisol level in the body leading to massive hair shedding. Magnesium present in millets control hair loss by reducing cortisol level in the body. Premature graying of hair is caused mainly due to oxidation of the tissues. The potent antioxidants in millets effectively prevent the corrosion of tissues, thereby reducing the possibility of premature graying.
- Menopause, the phase of life which no woman can escape from. Menopause brings with it a great deal of irritation and annoyance which is mainly due to hormonal imbalance. The intensity of menopausal symptoms experienced by every woman varies, depending upon her tolerance level and her other health and nutritional issues. Intake of food like ragi will not only provide nourishment but also prevents osteoporosis which is a very commonly faced problem during menopause. It also helps to keep mind and body calm.
- Due to its high level of magnesium, ragi is a great food for women who suffer from unbearable pain and cramps during their menstrual cycle.
- Research has shown that calcium in this millet helps to reduce irritability and nervousness to some extent. It has also been found to be particularly effective in alleviating symptoms associated with PMS (pre-menstrual stress).
- Celiac is a disease which damages the small intestine and interferes with the absorption of nutrients from food. People who suffer from this disease cannot tolerate gluten. This makes millet a perfect food for them since it is completely gluten-free. It is a wonderful grain alternative for people who are gluten-sensitive. Ragi malt is good for those who are allergic to milk and dairy products. It is one of the least allergic and most digestible grains.
- Consuming foods high in fibre also prevent the formation of gallstones, especially in women. Insoluble fibre speeds up intestinal transit time and reduces the secretion of bile acids that contribute to the gallstone formation.
- It also contains phosphorus which is conducive to control blood pressure. These minerals, along with the phyto-nutrients can help to build a formidable defense against all types of diseases and cancers.

*6. Kodo millet (**Paspalum scrobiculatum**)*

The kodo millet, also known as cow grass, rice grass, ditch millet, or Indian Crown grass, is an annual grain that is grown primarily in India. Kodo millet is a gluten free and vegan cereal. It originates in tropical Africa and it is estimated to have been domesticated in India 3000 years ago. It is grown mainly in Tamil Nadu and consumed by local people. Inspite of cultivating Kodo millet for 1000's of years, this millet is slowly dying out. Kodo millet is a nutritious grain and a good substitute to rice or wheat. Kodo millet has around 11 percent protein and the nutritional value of the protein has been found to be slightly better than that of foxtail millet, but comparable to that of other small millets. The presence of phospholipids, fibre contents and low oil content makes this grain a true nutraceutical, however, it is a minor crop as there is no consumer market for this millet. Kodo is grown mostly in organic way as it is a farmer friendly crop which can sustain heavy drought and marginal soils. In recent days, this grain has got a re-birth. It is also grown in the Philippines, Indonesia, Vietnam, Thailand, and in West Africa where it originates. It is grown as a minor crop in most of these areas, with the exception of the Deccan plateau in India where it is grown as a major food source. Kodo millet has large potential to provide nourishing food to subsistence farmers in Africa and elsewhere. Kodo millet prefers full light for optimal growth, but can tolerate some partial shading. Its ideal temperature for growth is 25–27°C. It requires four months until maturity and harvesting. Kodo millet has a shallow root system which may be ideal for intercropping. Kodo millet is propagated from seed, ideally in row planting instead of broadcast sowing.

Fig. 46: *Grains of Kodo millets (Paspalum scrobiculatum)*

The grain is composed of 11 percent of protein and 37–38 percent of fibre. Kodo millet contains 66.6 g of carbohydrates and 353 kcal per 100 g of grain, comparable to other millets.

It also contains 3.6 g of fat per 100 g. It provides minimal amounts of iron, at 0.5/100 mg, and minimal amounts of calcium, and 27/100 mg. Kodo millets also contain high amounts of polyphenols. An adequate fibre source helps combat the feeling of hunger.

Specialities of Kodo millet

- Kodo millet is very easy to digest, it contains a high amount of lecithin and is excellent for strengthening the nervous system.
- Kodo millets are rich in B-vitamins, especially niacin, B6 and folic acid, as well as the minerals, such as calcium, iron, potassium, magnesium and zinc.
- Kodo millets do not contain gluten and is good for people who are gluten intolerant.
- Regular consumption of kodo millet is very beneficial for post-menopausal women suffering from signs of cardiovascular disease, like high blood pressure and high cholesterol levels. The fibre content of the whole grain is very high.
- Five anti-diabetic compounds i.e., quercetin (the major one), ferulic acid, p-hydroxybenzoic acid, vanillic acid and syringic acid has been isolated from kodo millet. Quercetin prevents obesity.
- Kodo millet showed the highest phenolic content in the studies. It is consumed as a food by diabetic patients, since traditionally it was believed in reducing sugar level.

*7. Oat **(Avena Sativa)***

Oats are one of the most endowed grains and come among the top nutrient-rich grains in the world. The single largest area under oat in the world is in Russia, which is 30% of the global production area. Canada and Australia are second and third biggest, respectively. Within European agriculture, Poland, Spain, Finland and the Ukraine are the largest producers.

Oats require a moist, cool climate with a lot of rain and sunshine during the growing period and a lot of light and long days during the ripening period. As small grains need moisture during winter, spring and early summer, it is important that water be provided to the crop during this period. It does not require low temperatures, but not damages by low temperatures and frost during the early growth stages. Small grains in the germinating, seedling and vegetative stages are not sensitive to low temperatures and frost. Winter types require very low soil temperatures (3 to 4°C) for at least 10 days to 8 weeks for vernalisation and to reach their tillering potential

Oats are generally considered healthy due to their rich content of several essential nutrients. In a 100 gram serving, oats provide 389 calories and are an excellent source (20% or more of the Daily Value, DV) of protein (34% DV), dietary fibre (44% DV), several B–vitamins and numerous dietary minerals, especially manganese (23% DV). Oats contain 66% carbohydrates, 11% dietary fibre and 4% beta-glucans, 7% fat and 17% protein.

Fig. 47: *Grains of Oat (Avena Sativa)*

Specialities of Oats

- Daily consumption of oats over weeks lowers LDL ("bad") and total cholesterol, possibly reducing the risk of heart disease. One type of soluble fibre, beta-glucans, has been proven to lower cholesterol.
- Oats are the only cereal containing a globulin or legume-like protein, avenalin, as the major (80%) storage protein. Globulins are characterised by solubility in dilute saline as opposed to the more typical cereal proteins, such as gluten and zein, the prolamines (prolamins). The minor protein of oat is a prolamine, avenin. Oat's prolamins, named avenins, are similar to gliadins found in wheat, hordeins in barley and secalins in rye, which are collectively named gluten.
- Protein of oats is nearly equivalent in quality to the protein of soy, meat, milk and eggs. The protein content of the hull-less oat kernel (groat) ranges from 12 to 24%, the highest among cereals.
- Oats contain soluble fibre from whole oats (oat bran, oat flour and rolled oats), noting that 3.0 grams of soluble fibre daily from these foods may reduce the risk of heart disease.
- Oats, after corn (maize), have the highest lipid content of any cereal, e.g., greater than oats 10% and as high as 17% for some maize cultivars compared to about 2–3% for wheat and most other cereals. The polar lipid content of oats (about 8–17% glycolipid and 10–20% phospholipid or a total of about 33%) is greater than that of other cereals, since much of the lipid fraction is contained within the endosperm.
- Oats are considered one of the best remedies for "feeding" the nervous system, particularly in times of stress and in the case of nervous system weakness or exhaustion associated with depression.
- Oats are also a love potion, supporting sexual health and increasing libido by nourishing the endocrine system and regulating hormones, moistening glands and restoring nerve health.

- Oats contain carbohydrates in high amounts which is the most abundant nutrient in oats. Carbohydrates in oats are easily assimilated and also absorbed slowly in the body that is why they provide energy for several hours.
- Starch form other sugars like maltose, dextrose and glucose are derived. All these are assimilated easily into the body and releases energy with almost immediate effect.
    - **Fructose:** This is found in small amounts together with other carbohydrates. It does not require insulin before they are absorbed and utilized by the body cells. This property makes them highly tolerated by the diabetics.
    - **Mucilage:** This is also a form of carbohydrate present in oats that has the propensity to retain water. This property makes it very suitable for those with gastritis and colitis.
- Oats are considerably rich in minerals, as they contain iron and phosphorus in very high amounts. The abundance of these two minerals in oats surpasses that of meat.
- Vitamins are not left out in oats, as they are contained in reasonable amounts. One of the vitamins that stand out in oats is vitamin B1, as is attributed to some of its medicinal benefits.
- Oats contain proteins in abundant amounts, representing about 17% of its total content. The proteins contain all the essential amino acids, but not in the required proportions. Moreover, oats contain a few amounts of lysine and threonine, and have an excess amount of methionine. Likewise, legumes such as beans or lentils are very rich in lysine, but lacks methionine. Due to these deficiencies on both hands, the combination of oats with legumes is highly recommended in order to supply the body with complete proteins.
- Fats are also found in oats made up of about 80% unsaturated fatty acids and only about 20% saturated acids. This dominance of unsaturated fatty acids in oats, makes it to have a regulating effect on the synthesis of cholesterol.
- Fiber in oats is important component which mainly found in its outer layer and its bran. It contains beta-glucon which is a soluble substance derived from cellulose. It reduces cholesterol levels due to its actions in absorbing and removing bile acids, which are the major components of cholesterol. Apart from its cholesterol lowering effect, it also acts as a mild laxative.
- Oats helps in lowering cholesterol level and also has some benefits to the nervous system. Another substance present in oats is Avenasterol which is similar to cholesterol and plays an interesting role by interfering with the absorption of main cholesterol.
- Oats reduces body's cholesterol level and eliminate the bile salts in the intestine which forms cholesterol as well as reduce the absorption of fats by the body.
- Several substances present in oats have been proved to regulate the level cholesterol, an important cause of arteriosclerosis and hypertension. Thus, the regular consumption of oats is highly recommended in cases of both disorders.

- Oats, which contain gliadin in reasonable amounts is well tolerated by celiac disease patients. Thus, the consumption of oats can be of immense benefits to those suffering from this disease.
- The presence of some important substances in oats such as fatty acids, phosphorus, lecithin, vitamin B1 and glucose, enhances the functions of the nervous system. Additionally, presence of an important substance known as avenine in oats has a sedating effect on the nervous system. The regular consumption of oats is highly recommended in some cases like nervousness, insomnia, fatigue and depression.
- Oats contains very high carbohydrate content, being dominated by fructose. Moreover, beta-glucan a substance found in oats bran, has been proved to greatly improve the tolerance of glucose released from starch during digestion of food materials.
- The mucilage present in oats possess a soothing effect, as well as ease digestion process. The consumption of oats in combination with either milk or broth is highly recommended in cases of gastroduodenal, ulcer, gastritis, colitis and diverticulosis.

*8. Buck Wheat (**Fagopyrum esculentum**)*

Buck wheat, not like wheat, is reffered to as pseudo-cereal and known as power-house of nutrients. Buck wheat was domesticated in South-East Asia and from there spread to Central Asia and Tibet and than to Middle East and Europe. Whole buck wheats scored highest on their ability to satisfy hunger and is great food for eating in the cold winter months. Its production has declined in 20th century. Grains of Buck wheat are rich in carbohydrates and have fewer calories in comparison to wheat or barley. It is free of saturated fat and cholesterol and rich in dietary fibre and gluten free protein. This combination plays an important role in suppressing appetite, controlling blood sugar, facilitating proper digestion and building lean muscle mass. Buckwheat is a great source of plant-based protein and contains twelve amino acids - the "building blocks of protein" that support energy, growth and muscle synthesis. Its protein content is higher than all foodgrains except quinoa and beans.

Fig. 48: *Grains of Buck Wheat (Fagopyrum esculentum)*

When buckwheat or any whole grain is refined, its bran and germ are removed. Although these two parts make up only 15–17 percent of the grain's weight, they contain 83 percent of its phenolics. One type of phyto-nutrient especially abundant in whole grains of buckwheat are plant lignans, which are converted by friendly flora in our intestines into mammalian lignans, including one called enterolactone that is thought to protect against breast and other hormone-dependent cancers as well as heart disease.

Specialities of buck wheat

- Buck wheat is rich in flavonoids, particularly *rutin* that protects against disease by extending the action of vitamin-C and acting as antioxidants. The high level of *rutin* is extracted from the leaves of buckwheat plants for medicine to treat high blood pressure.
- Buck wheat is rich in flavonoid compounds which help in maintaining blood flow, keep platelets from clotting excessively (platelets are compounds in blood that, when triggered, clump together, thus preventing excessive blood loss, and protect LDL from free radical oxidation into potentially harmful cholesterol oxides). It lowers blood sugars more slowly than rice or wheat products.
- Buckwheat is a good source of magnesium which is instrumental in lowering blood pressure by relaxing the blood vessels without the use of harmful chemicals. This mineral relaxes blood vessels, improving blood flow and nutrient delivery, while lowering blood pressure, the perfect combination for a healthy cardiovascular system. Magnesium is also beneficial for bone and dental health and production of energy. Magnesium also facilitates the absorption of calcium by the body and prevents the development of osteoporosis.
- Manganese present in abundant in buckwheat ensures healthy bone structure by forming essential enzymes for building bones and acting as a co-enzyme for assisting metabolic progression in the human body. It is also actively involved in the formation of connective tissues, absorption of calcium and metabolism of fat and sugar.
- Its magnesium also lower the risk of diabetes in women, thus, it is more helpful in the management of diabetes. The component in buckwheat responsible for its blood glucose-lowering effects appears to be chiroinositol, which plays a significant role in glucose metabolism and cell signaling. It improves micro-vascular integrity and circulation in diabetics, thus preventing the damage of nerves and muscle cells and loss of kidney function. Being low in calories and fat free, buckwheat is an ideal food for diabetic patients.
- Buckwheat is rich in insoluble fibre which helps women to avoid gallstones, increases insulin sensitivity, but lowers the secretion of bile acids and blood sugar. Fibre not only speeds intestinal transit time (how quickly food moves through the intestines), but reduces the secretion of bile acids

(excessive amounts contribute to gallstone formation), increases insulin sensitivity and lowers triglycerides (blood fats).

- Continouos consumption of buckwheat slowed progression of atherosclerosis, the build-up of plaque that narrows the vessels through which blood flows, and less progression in stenosis, the narrowing of the diameter of arterial passage-ways.
- Due to its neutral thermal nature and sweet flavour, buckwheat cleans and strengthens the intestines and improves appetite. It is effective for treating dysentery and chronic diarrhea.
- Being rich in vitamins-B, particularly niacin, folate and vitamin B6, buckwheat is highly beneficial for the cardiovascular health. These vitamins reduce the concentration of cholesterol in the blood. Niacin causes an increase in high density lipoproteins (HDL) which further enhance the blood vessel strength and cholesterol removal.
- Buckwheat contains high quality proteins which remove the plaque forming triglycerides and low density lipoproteins (LDL). Thus buckwheat is highly beneficial for people with weak heart functions and other cardiovascular problems.
- Buckwheat reduces the risk of breast cancer. The plant lignans are converted into mammalian lignans (called enterolactone) in the intestines. It is believed to protect against breast cancer and other forms of cancer related to hormones.
- Buckwheat contains proteins of a very high quality which offers several health benefits. This protein in combination with amino acids helps in lowering cholesterol levels besides being a key factor in preventing diabetes and obesity. In addition, buckwheat proteins reduce hypertension by reducing the activity of angiotensin converting enzyme.
- High protein buckwheat flour is being studied for possible use in foods to reduce plasma cholesterol, body fat, and cholesterol gallstones. Being high in insoluble fibre, buckwheat can help in preventing gallstones.
- Buckwheat has an important role to play in skin health.
- Whole grain buckwheat can be very helpful in weight loss.

*9. Sorghum (**Hordium vulgare**)*

Sorghum is the fifth most important cereal crop in the world after wheat, rice, maize and barley. It is grown largely in the arid and semi-arid parts of the world, due to its feature of being extremely drought tolerant. It probably originated in Ethiopia and has spread to other parts of Africa, India, South-east Asia, Australia and the United States.

More than 55 percent of the total production is consumed as a food for human being and around 33 percent is used as fodder. The nutritional value of sorghum is same as of that of corn and that is why it is gaining importance as livestock feed. Sorghum is also used for production of ethanol, alcohol, starch, adhesives and paper other than being used as food and feed. The popularity of the crop is clear from the fact that it is cultivated in more than 100 countries in the world. USA rank first accounting up to 17 percent of the world's total production. Nigeria is second and India is on 3rd position among major producer countries. Regarding the consumption pattern, Nigeria being at the top of the list. USA contributing around 88% of the world's total exports. Argentina, Australia, Nigeria, China and India are the other important exporters of this cereal grain.

Fig. 49: *Grains of Sorghum (Hordium vulgare)*

India has ever been among the major producers of sorghum in the world and has been able to maintain its position among the top three producers of the crop. The major states in the country where this cereal grain is produced are Maharashtra, Karnataka, Gujarat, Madhya Pradesh, Andhra Pradesh, Rajasthan and Uttar Pradesh. Maharashtra produces the maximum sorghum in India, production being supported by the districts Solapur, Pune, Bijapur, Nanded, Akola and Maboobnagar.

It can tolerate high temperature throughout its life cycle better than any other crop. Sorghum requires about 26–30°C temperature for good growth. The minimum temperature for the germination of sorghum seed is 7 to 10°C. Sorghum requires warm climate, but it can be grown under wide range of agro-climatic conditions. It is best adapted to areas having an average annual rainfall between 45 to 65 cm.

Specialities of Sorghum

- Sorghum is best source of good dietary fibre which greatly improves the functionality of the digestive system and treats conditions like gas, diarrhea, bloating, constipation and stomach ache. Dietary fibre is a bulking agent that adds bulk to the stool and helps it pass smoothly through the digestive tract.
- Fiber prevent cramping, bloating, constipation, stomach aches, excess gas, and diarrhea. Furthermore, excess amounts of fibre in the body helps to scrape off dangerous cholesterol (LDL), which helps to improve heart health and protect our body from conditions like atherosclerosis, heart attacks and strokes.
- Bran layer of sorghum contains many cancer-fighting and free-radical eliminating antioxidants that are essential for the body. Some of these antioxidants are exclusive, thus, sorghum becomes an effective anti-carcinogenic. Antioxidants

are the beneficial compounds that neutralize and eliminate free radicals in the body, which often cause healthy cells in the body to mutate into cancer cells.

- Shorghum grains are gluten free. Gluten is a widely used ingredient that makes the life of celiac disease victims quite hard. Sorghum can be eaten without suffering from inflammation, gastrointestinal damage and nausea.
- Sorghum helps in control of diabetes. Excess carbohydrates usually break down simple sugars and increase the body's glucose levels that often leads to diabetes. Sorghum is rich in tannin, contains enzymes that prohibit starch absorption by the body. This property helps to regulate insulin and glucose levels. A balance in the levels ensures that diabetics don't have to suffer sudden spikes and drops in glucose levels, thereby preventing diabetic shock and other complications.
- Magnesium is found in high quantities in sorghum, which increases calcium absorption that helps to maintain calcium levels. These two minerals are also integral to the development of bone tissue and speeds up the healing of damaged or ageing bones. This can prevent conditions like osteoporosis and arthritis, thereby keeping us active and healthy into our old age.
- Sorghum improves circulation as it is rich in many minerals, including iron and copper. Copper boosts the uptake and absorption of iron and ultimately regulate the iron deficiency (anemia) in the body. With enough iron and copper in our system, red blood cell development is increased, thereby boosting circulation of the blood, stimulating cellular growth and repair, and increasing the hair growth of the scalp, while also boosting energy levels throughout the body.
- Sorghum works as energy booster. It contains Niacin, also known as Vitamin B_3, which transforms food into usable energy and fuel for the body. Vitamin B_3 metabolizes nutrients and helps to maintain energy levels throughout the day.
- Sorghum contains many cancer-fighting antioxidants that can help to treat cancers, especially melanoma (special type of skin cancer), which is characterized by pigment-containing cell growth on the body. The tanning in Sorghum may have restrictive effects on growth of melanoma cells.
- Sorghum is crammed with essential nutrients that are not only necessary for development, but also help keep the body free from diseases and maintain its health.
- Sorghum is rich in manganese which helps to function the thyroid glands properly which results in losing weight, appetite, efficient organ systems and metabolism.
- Phosphorus helps to maintain the response of emotions, neurons and hormones. The phosphorus deficiency is associated with the decrease in cognitive power and age related neuro-degenerative disorders, such as dementia and Alzheimer's disease.
- Vitamin B6 has a huge role in the production of serotonin and GABA neuro-transmitters in the brain which controls the mood. Vitamin B6 is essential for to hinder pain, depression, anxiety and fatigue. This shows that Vitamin B6 helps to prevent mood disorders and raise mood.

10. Freekeh

Freekeh is an ancient Middle Eastern dish and name is given to any grain mainly wheat which is harvested young and green (immature) and then roasted. Freekeh is truly worthy of being called a super food. It is similar to brown rice, but has up to 3 times the amount of fibre and protein, making this grain a true super grain. It is rich in iron. Freekeh retains higher amounts of protein, vitamins and minerals. It is also packe with fibre which is more than 2 to 4 times more than brown rice. It offers much more than quinoa. It is not gluten free as it is wheat, but, it carries amazing digestive and other benefits.

Fig. 50: *Grains of Freekeh*

It has double amount of fibre and protein that quinoa does. This may results in weight loss for those adding it to a healthy life style. For those living with diabetes, Freekeh is also the better option – it ranks at a low 43 on the glycemic index, where as quinoa is higher at 54. Freekeh protect against weight gain. In stomach, freekeh acts as a prebiotic-providing sustenance for the good bacteria that aid in digestion.

Specialities

- It can ease a number of common digestive complaints including diarrhea and constipation due to higher fibre content.
- It aid weight control management.
- It contains loads of antioxidants (zeaxanthin) which can have possitively impact on eye health. These are known to prevent macular degeneration which typically occurs with age.
- It aids muscular growth and endurance naturally due to glutamic acid for body builders and atheltes.
- It is ideal food for maintaing blood sugar levels due to low glycemic index.

11. Black barley

Barley, the oldest cultivated grains, has been modified to produce greater yield and economic returns. It is a variety of barley (contains gluten) originates from Ethiopia. It offers all the nutritional benefits of modern barley. Its grains are black because the outer bran is left intact (rather than hulled) and rich in anthocyanin pigment, which have health promoting properties. This barley has more of a chewy and nutty bite than the more commonly used barley, where the outer hull has been removed. The appearance of grains like mahogany colour develops a rich, glossy sheen when cooked black barley has a more vibrantly nutty taste than traditional barley. Black barley take a bit longer to cook than hulled barly. By soaking grains in water overnight reduces cooking time. As a 100 percent wholegrain, black barley is rich in nutritional values, particularly high in iron, zinc and heart-healthy fibres. Like barley, black barley can be added to soups, salads, stuffings, pilapse, and more used on its own or in combination with other grains.

Fig. 51: *Grains of black barley*

12. Farro

Farro is Itelian name which is an ancient variety of wheat. Ancient varieties of wheat are nutritionally similar to modern day wheats, however, generally higher in protein and fibre, particularly insoluble fibre which is important for digestive well-being. Farro is similar in texture to rice and has a nutty flavour, making it great grain, but not gluten free.

Fig. 52: *Grains of Farro*

13. Kamut

Kamut is an ancient Egyptian grain similar to wheat which is rich in amino acids, proteins, minerals and selenium. It is also filled with antioxidants that boost and protect the immune system. This power-packed grain is ideal for backed foods. It can help lower bad cholesterol and rise good cholesterol as it is rich in essential fatty acids. Since its grains are little tough, it may be soaked in water over night before cooking.

Fig. 53: *Grains of Kamut:*

14. Brown rice

Brown rice, also called hulled rice, is rice that has either not been milled at all or only been partly milled. This means that only the kernel (or the hull of the rice) is removed. Brown rice is a healthier option than white rice. One cup of cooked brown rice (of the long grain variety) will have about 216 calories. One cup of brown rice provides 80% of our daily manganese requirements. Brown rice is a natural wholesome food rich in essential minerals, such as manganese, iron, zinc, phosphorous, calcium, selenium, magnesium and potassium. Vitamin wealth includes vitamin B1 (thiamine), vitamin B2 (riboflavin), vitamin B3 (niacin), vitamin B6, folate, vitamin E (alpha-tocopherol) and vitamin K. It is a source of protein and adds good amount of fibre content to our diet. Along with this, it is also a provider of health-supportive vital fatty acids. Brown rice is simply delicious and a fantastic staple for both vegetarian and vegan diets. Brown rice can be used as a white rice alternative in most vegetarian recipes and provides a full, rich and somewhat nutty flavour. Brown rice flour can be used for vegetarian pancakes, breads and other baked goods. All in all, brown rice is clearly the healthy choice.

Fig. 54: *Grains of brown rice*

Specialities of brown rice

- Brown rice is a good source of selenium which reduces the risk for developing common illnesses, such as cancer, heart disease and arthritis. Selenium also plays a role in the antioxidant process, can destroy cancer cells and even repair DNA and regulate the thyroid hormone metabolism and immune system function.
- It is high in Manganese which helps the body to synthesize fats and plays an important part in fighting free radicals. Manganese is part of a compound known as superoxide dismutase, an antioxidant that prevents damage from free radicals created during the energy production process. Manganese is also important for deriving energy from protein and carbohydrates, and plays a key role in the synthesis of fatty acids. Manganese also beneficial for our nervous and reproductive systems.
- Brown rice is a good source of magnesium, a mineral that is essential to bone health. Most of the magnesium in the body is stored in the bones, and to reap the other health benefits, brown rice is reommended to eat each week.
- The oil in brown rice has been shown to lower levels of LDL cholesterol, also known as the bad cholesterol, by up to seven percent. At the same time, a diet high in whole grains can increase the level of HDL cholesterol, or the good cholesterol.

- Consuming atleast two servings of brown rice per week can lower the risk of developing diabetes. They found that by eating just 50 grams of brown rice a day, the risk of type 2 diabetes can be lowered by 16 percent, while other whole grains, such as barley and whole wheat, can lower the risk by 36 percent.
- Brown rice supplies 14 percent of the recommended daily value for fibre, which tends to latch onto the chemicals that cause cancer and steer them away from the cells in the colon and breasts, preventing cancer from developing in those areas. Fiber also has many other health benefits including promoting cardiovascular health. Thus, it is an important nutrient that protects against colon cancer and breast cancer.
- Brown rice makes the tummy feel full which translates to smaller meal portions. It ultimately promotes weight loss.
- Brown rice is considered a whole grain as it has not lost its "wholeness" through the refinement process. Whole grains are proven to reduce the buildup of arterial plaque and reduce the risk of heart disease and high cholesterol.
- It is rich in anti-oxidants which is one of the best kept secrets regarding brown rice. The antioxidant capacity of brown rice is right up there with these super stars.
- Brown rice is high in fibre and on top of the list for foods that can help to prevent colon cancer. This can be attributed to the high levels of fibre naturally contained in brown rice. These fibres attach to substances that cause cancer as well as to toxins in the body, thus eliminating them and keeping them from attaching to the colon wall.
- Brown rice helps stabilize blood sugar levels; therefore, it is an excellent food choice for those suffering from diabetes. Studies reveal that those who consume one half cup of brown rice daily reduce their risks of developing diabetes by 60 percent. On the other hand, those who consume white rice regularly increase their chances of developing diabetes one hundred-fold.
- It is considered as perfect Baby Food. Brown rice cereal or brown rice itself is the perfect baby's first food due to the dense natural nutrition and fibre it contains. This is a much better choice than refined white rice cereal products as rapidly growing babies and toddlers require nutrient rich diets help to maintain rapid growth cycles.
- Brown rice is good source of phytonutrients. Phyto-nutrients are compounds naturally found in plants that have anti-inflammatory properties and tend to act as an antioxidant. Brown rice is a great source of these plant compounds, especially phenolics. Infact, research has shown that brown rice contains almost as much phenolics as fruits and vegetables.
- It reduces childhood asthma which is a very common condition among children and causes many children to miss numerous days of school. But children who eat plenty of whole grains along with fish can lower their risk of developing asthma by 50 percent.

Overall benefits of super grains

- Reduces joint inflammation.
- Prevent hair, skin and nail disorders.
- Lower blood cholesterol.
- Reduce liver fat.
- Protect kidneys.
- Reduce bladder irritation.
- Promote relaxation.
- Stress and depression reduction.
- Relieving migraine headaches.
- Assisting the immune system.
- Reduces the risk of artery and heart spasms.
- Enhance calcium absorption in the body.
- Promote collagen formation.
- Enhances antibody, hormone and enzyme production.
- Maintain transmission of signals between the nerve cells and the brain, and alertness.
- Promote memory improvement.
- Smoothen digestive and intestinal tract functioning.
- Muscle coordination.
- Mental vigour.
- Manufacture of other essential biochemical components.

GLOSSARY

- **Abscisic acid:** Abscisic acid is a plant hormone that inhibits growth, causes abscission of leaves, induces dormancy, closes stomata and triggers other phenomena in response to adverse conditions.
- **Abscission:** Abscission is the normal separation of a leaf, fruit or flower from a plant or the dropping of leaves, flowers and fruits. Abscisic acid is the plant hormone involved in abscission.
- **Abscission layer**: Specialized cells, usually at the base of a leaf stalk or fruit stem, that trigger both the separation of the leaf or fruit and the development of scar tissue to protect the plant.
- **Absorption**: The intake of water and other materials through root or leaf cells.
- **Acid precipitation:** Any form of precipitation (rain, snow, hail or fog) whose acidity has been increased through the uptake of acid pollutants from the air.
- **Acid soil:** Soil with pH below 7 on a pH scale of 1 to 14. The lower the pH, the more acid the soil.
- **Active ingredient**: The chemical in a pesticide formulation that actually kills the target pest.
- **Acaricide:** A substance that kills mites or ticks.
- **Actinomorphic:** This term is used for a flower which means they can be divided into 3 or more identical sectors (related to each other by rotation about the centre of the flower). Typically, each sector might contain one tepal, or one petal and one sepal and so on.
- **Acuminate:** Tapering gradually into a point.
- **Actinomycetes:** A group of micro-organisms, intermediate between bacteria and true fungi that usually produce a characteristic branched mycelium. These organisms are responsible for the earthy smell of compost.
- **Adventitious bud:** A bud in on usual place on a plant, often on an internode. May be the result of injury. Suckers and water sprouts usually grow from adventitious buds.
- **Adventitious root:** A root in an unusual place, often where a branch contacts soil or damp material. A plant cannot be reproduced from cuttings or layering unless adventitious roots develop.
- **Adaxial:** The parts of a flower nearest to the axis of the branch.
- **Aeration:** The process by which oxygen-deficient air in soil is replaced by air from the atmosphere. Aeration can be enhanced by hoeing or other cultural

operations. Mechanically loosening or puncturing soil to increase permeability to water and air.

- **Aerobic:** Active in the presence of free oxygen (requiring oxygen). A condition identified by the presence of oxygen. An adjective describing an organism that can live only in the presence of oxygen (i.e., an aerobic organism).
- **Aggregate soil:** Many soil particles held in a single mass or cluster such as a clod, crumb, block or prism.
- **Agricultural wastes:** Wastes normally associated with the production and processing of food and fibre products on farms, feed lots, ranches and ranges. Waste produced as a result of various agricultural operations. It includes animal manure, crop residues and dead animals manure and other wastes from farms, poultry houses and slaughter houses; harvest waste; fertilizer run-off from fields; pesticides that enter into water, air or soils; and silt drained from fields.
- **Agricultural pollution:** Liquid and solid wastes from all types of farming activities, including run-off from pesticide and fertilizer use and from feed lots; erosion and dust from plowing animal manure and carcasses; and crop residues and debris.
- **Agro-forestry**: Land management which combines agricultural crops with tree crops and forest plants and/or animals simultaneously or sequentially and applies management practices which are compatible with the cultural patterns of the local population. Agroforestry is a land use system in which woody perennials are grown with agricultural crops (together with other land uses, like animal production).
- **Agrology**: Agrology is a branch of soil science that studies the soil used in producing crops.
- **Algae**: Algae are simple photosynthetic organisms that belong to the kingdom Protista. Most algae are aquatic; seaweeds are algae. Some algae are unicellular, while others are multi-cellular.
- **Alkaline soil**: Soil with pH above 7 on a pH scale of 1 to 14. The higher the reading, the more alkaline the soil.
- **Allelopathy:** The excretion by some plants of compounds from their leaves and/or roots that inhibit the growth of other plants.
- **Alien plants**: An alien plants that are not native to a place; it came from another place.
- **Anion:** A negatively charged ion. Plant nutrient examples include nitrate (NO3-), phosphate (H2PO4-) and sulfate (SO4-).
- **Anther:** The pollen-bearing part of a flower's male sexual organ. The filament supports the anther; together they are referred to as the stamen.
- **Air pollutants:** Substances in the air that could, at high enough concentrations, harm Human beings, animals, vegetation or material. Air pollutants may thus include forms of matter of almost any natural or artificial composition capable of being airborne. They may consist of solid particles, liquid droplets or of

being airborne. They may consist of solid particles, liquid droplets or gases, or combinations of these forms.

- **Air pollution:** The presence of contaminant or pollutant substances in the air that do not disperse properly and that interferes with human health or welfare, or produce other harmful environmental effects.
- **Air pollution Index:** Quantitative measure that describes ambient air quality. The index is obtained by combining figures for various air pollutants into a single measurement.
- **Airborne disease:** Disease that is generally transmitted by esopharyngeal discharges and by respiratory secretions, through coughing and sneezing, though it may also be conveyed through close contact. Respiratory diseases include the common childhood infections, measles, whooping cough, chickenpox, mumps, diphtheria and acute sore throat, as well as diseases of the respiratory tract, influenza and other acute viral infections, pneumonia, and pulmonary tuberculosis (WHO, 1992).
- **Aquaculture:** Rearing or raising under controlled conditions of aquatic products such as fish, oysters, seaweeds and other aquatic resources in sea, lakes, rivers and swamps and other bodies of water. Examples are fish-pond, fish-pen, and sea-weed cultures.
- **Atmosphere:** Mass of air surrounding the earth, composed largely of oxygen and nitrogen.
- **Atmospheric absorption:** Absorption by the earth's atmosphere of most of the X-rays and ultraviolet and infrared radiation emitted by the sun, except visible light. It prevents the earth's surface from becoming too hot.
- **Average annual rainfall**: Specified for a place; it is the mean of the annual amounts over the period for which rainfall statistics are available.
- **Agronomic nutrient rate:** Amount of supplemental nutrient required by a crop for a realistic yield goal (yield potential), after all the soil, water, plant, and air credits are considered. Agronomic rates consider nutrient credits from all soil tests, legumes, manure residuals and any other nutrient source.
- **Ammonium (NH_4^+):** A form of nitrogen that is available to plants and is derived from fertilizer addition and soil organic matter mineralization.
- **Ammonium nitrate solution:** Non-pressure solution of ammonium nitrate in water usually standardized to 20 percent nitrogen used for direct application or for making multinutrient liquid fertilizer. Analysis is 20-0-0.
- **Ammonium Nitrate Limestone:** A combination of ammonium nitrate and limes stone to form a product containing 20.5 percent nitrogen.
- **Ammoniation:** A process wherein ammonia (anhydrous, aqua or a solution containing ammonia and other forms of nitrogen) is used to treat superphosphate to form ammoniated superphosphate, or to treat a mixture of fertilzer ingredients (including superphosphate) in the manufacture of a multiple-nutrient fertilizer.
- **Amorphous:** Having no definite shape or structure.
- **Anaerobic:** A condition identified by the absence of oxygen.

- **Anhydrous ammonia (NH3):** Fertilizer in pressurized gas form, made by reacting nitrogen in air with hydrogen gas (H2) under high temperature and pressure in the presence of a catalyst.
- **Anion:** An ion with a negative charge. Common soil anions are chloride, nitrate, sulfate and bicarbonate.
- **Anion exchange capacity:** The sum total of exchangeable anions that a soil can adsorb at a specific pH. Expressed as centimoles of charge per kilogram (cmolc/kg) of soil or milli-equivalents per 100 g of soil (meq/100 g of soil).
- **Application rate:** The weight or volume of a fertilizer, soil amendment, or pesticide applied per unit area.
- **Aqua ammonia:** A 20% N solution made by dissolving anhydrous ammonia in water.
- **Available nutrient:** A nutrient in a form that a plant can absorb.
- **Amino acid:** Building blocks of proteins. Plants have 22 types of amino acids.
- **Anaerobic:** Not requiring oxygen.
- **Anhydrous Ammonia (NH_3):** A common form of nitrogen fertilizer used by wheat growers and other farmers. A gas containing approximately 82 percent nitrogen. Under pressure, ammonia gas is changed to a liquid and is stored and transported in this form'
- **Anther:** Male reproductive structure that produces pollen in plants.
- **Angiosperm:** Angiosperms (meaning "covered seed") are flowering plants. They produce seeds enclosed in fruit (an ovary). They are the dominant type of plant today; there are over 250,000 species. Their flowers are used in reproduction. Angiosperms evolved about 145 million years ago, during the late Jurassic period, and were eaten by dinosaurs. They became the dominant land plants about 100 million years ago (edging out conifers, a type of gymnosperm). Angiosperms are divided into the monocots (like corn) and dicots (like beans).
- **Anthophyta**: Anthophyta are flowering plants, the largest group of plants (which includes the grasses). The flowers are used in reproduction. They evolved during the Cretaceous period.
- **Androecium:** A collective name for the stamens.
- **Androgynous:** It is the situation when male and female flowers on the plant are mixed in a spike or head.
- **Androgynophore:** A column composed of an ovary on a long stalk to which the stamens are attached. The staminal filaments form a connate tube and the stamens continue past the ovary and the free parts hold the anthers (Passifloraceae).
- **Andromonoecious**: It is the situation when male and bi-sexual flowers are developed on different individuals.
- **Anthesis**: The time of flowering.
- **Annual weed:** Weeds that complete their vegetative and reproductive life cycle (from growing to seed formation) within one growing season or in a year.

- **Aphids:** Small insects that feed by sucking the sap from plants.
- **Apical meristem:** The apical meristem consists of meristematic cells located at the tip (apex) of a root or shoot.
- **Apical dominance**: Apical dominance is the phenomenon in which a terminal (end) bud inhibits the development of lateral (side) buds. The inhibition of lateral bud growth by the presence of the hormone auxin in a plant's terminal bud. Removing the growing tip removes auxin and promotes lateral bud resulting subsequent branching.
- **Apogeotropic roots:** Apogeotropic roots grow upwards to the soil surface (other roots grow downwards), emerging from the soil and growing upwards.
- **Apomixis**: Apomixis is a type of reproduction in which a plant produces seeds without fertilization.
- **Asexual reproduction:** Reproduction or propagation that does not involve the union of gametes.
- **Arable**: Suitable land for growing crop plants.
- **Arid**: An arid area is dry and hot, with little rainfall and few plants.
- **ATP:** ATP (Adenosine triphosphate), is a high-energy molecule used for energy storage by organisms. ATP is produced in the cristae of mitochondria and chloroplasts.
- **Available phosphate:** The correct chemical name for P_2O_5 is phosphorus pentoxide, which is never found in fertilzer, but historically, has been an accepted term for expressing content of phosphorus as available phosphate. The traditional term has been available phosphoric acid. It is chemically, phosphoric acid (H_3PO_4).
- **Auxin:** Auxins are best known and most important plant growth hormones. Most abundantly produced actively on growing tips and induce phototropism, apical dominance, cell elongation and many other reactions. Generally stimulates growth by cell division in the tip region and by cell elongation lower down the shoot.
- **Axis:** The axis of a plant runs through the middle of it, e.g., the stem of a plant or the rachis of a compound leaf.
- **Axillary bud:** The bud that develops in the axil (the angle between the stem and the leaf) of a plant.
- **Axil:** The axil of a plant is the angle between the upper side of the stem and a leaf, branch, or petiole. In flowering plants, the bud develop in the axil of a leaf.
- **Band seeding:** Sowing seed and fertilizer in the one operation, but through separate outlets.
- **Bone Meal**: Raw bone meal is cooked bones ground to a meal without any of the gelatine or glue removed. Steamed bone meal has been steamed under pressure to dissolve part of the gelatine.
- **Bio-control/Biological control:** A management practice in which various predatory or parasitic organisims are used to control pests and pathogens. They are usually part of an integrated pest management strategy for crops, especially vegetables.

- **Biotic:** Biological factors such as insects, disease pathogens, nematodes and weeds that affect other living organisms.
- **Border row:** The recommended number of rows of male parental lines to be grown on all the sides of the hybrid seed production.
- **Broadcast:** It is a method of spreading fertilizer / seeds / weedicides on the soil surface across the entire width of a cropped or planted area.
- **Biodiversity:** The variability among living organisms from all sources, including terrestrial, marine, and other aquatic ecosystems and the ecological complexes of which they are part. It includes diversity within species (genetic diversity), between species (species diversity) and between ecosystems (ecosystem diversity).
- **Biological diversity:** This refers to the total variety of life. It includes the range of plants and animals, the genetic variety within each species, populations and communities and the variety of habitats in which they live.
- **Bio-energy:** Energy that is derived from biological matter (i.e., from plants and animals), but which has not undergone a geological process (cf. fossil fuels). Carriers of bio-energy may be solid (e.g., wood, straw), liquid (e.g., bio-diesel, bio-ethanol) or gaseous (e.g., methane).
- **Biomass:** This is all living matter plus all matter that was living, but is in the process of decay. When the term biomass is used in an agricultural context, it may carry a more restricted meaning. It may mean plant matter that can be used as a source of energy. In an agricultural context, therefore, biomass may includes wood, crops, algae, agricultural and forest residues, sewage, manure, industrial by-products and municipal solid waste.
- **Biosecurity:** The protection of plants, animals, the natural environment and the economy from pests, disease and weeds.
- **Biosphere:** The regions of the earth inhabited by living organisms.
- **Biota:** The total of the living organisms in any particular area.
- **Bioturbation:** The mixing of the soil by living organisms such as worms.
- **Broad-leaved weed:** The leaves are "broad" as opposed to the "narrow" leaves of grasses. Also called 'dicots' having two seed leaves, while grasses are 'monocots' having one seed leaf.
- **Celiac disease:** Celiac is a disease which damages the small intestine, interferes with the absorption of nutrients from food and caused due to digestive problem. People who suffer from this disease cannot tolerate gluten as this disease has a severe allergy to gluten and gluten promote this disease.
- **Chlorosis**: A condition caused by low chlorophyll content and characterized by yellowing of the leaves of affected plants. Often a symptom of nutrient deficiency.
- **Climate change:** The long term change in the earth's climate, recently by an increase in surface temperature of the earth due to both natural and human factors.
- **Clone:** A group of individuals of common ancestry that have been propagated vegetatively (a sexually) usually by cutting or natural multiplication of bulbs or tubers.

- **Conservation tillage:** Also known as zero tillage, is a way of growing crops from year to year without disturbing the soil / cultivating the soil usually with tractor-drawn implements. Any tillage and planting system that covers 30% or more of the soil surface with crop residue after planting to reduce soil erosion by water.
- **Cover crop:** Cover crops are primarily planted not to be harvested for food but to reduce soil erosion, control weeds and improve soil quality. They are usually ploughed or tilled under before the next food crop is planted, in which cases the "cover crop" is used as a soil amendment and is synonymous with "green manure crop."
- **Contact herbicide:** A herbicide that kills only the tissues of the plants which come in its contacts.
- **Cross-breeding:** The mating of animals / or plants of different breeds, both of which are purebred, in order to combine the best characteristics of the two breeds. The pogeny, known as crossbreds, possess hybrid vigour. They are sometimes crossed with a male of a third breed to introduce further qualities.
- **Cross-pollination:** The transfer of pollen from an anther of one flower to the stigma of a second flower of a different cultivar.
- **Denitrification:** It is the process of changing nitrate or nitrite forms of nitrogen to the gaseous forms.
- **Deresination:** Removal of pericarp and the saponins from quinoa seeds by mechanical or chemical means called deresination. It does not affect the mineral content of the seed.
- **Deciduous:** Plants that shed their leaves annually. It can also refer to the process of annual bark shed.
- **Dew-point:** The temperature at which a mixture of air and water vapour will reach 100 per cent relative humidity or at which the mixture starts to condense.
- **Dicotyledon:** The class of plants with embryos that have two cotyledons. For example, beans and clover.
- **Dormancy:** The state of inhibited growth of viable seeds, even in the presence of otherwise favourable germinating conditions due to external factors.
- **Drought:** Meteorological drought when rainfall is within the lowest 10 per cent of all totals for at least three months.
- **Duplex soil:** A soil that has a sharp change in soil texture between the A and B horizons.
- **Electrical conductivity:** A measure of soil salinity. It measures the concentration of soluble salts in the soil solution, which plant roots absorb. Electrical conductivity (or EC) is expressed in units of decisiemens per metre (dS/m).
- **Electrolyte:** This term is used in context to salty solution. Electrolyte solutions are normally formed when a salt is placed into a solvent such as water and the individual components dissociate due to the thermodynamic interactions between solvent and solute molecules, in a process called solvation.

- **El niño:** The name given to warming of coastal waters off Peru around Christmas. Occasionally, this warming is exceptionally strong. It is these occasional extremes (once every 5–9 years) that are called El niño episodes. This strong warming of the eastern Pacific is linked to negative phases of the southern oscillation. Through this connection, episodes are generally periods of drought over eastern Australia.
- **Emergent vegetation:** A plant or plants rising above the water level, as opposed to floating or submerged vegetation.
- **Emasculation:** Removal of stamenes / male organ of a complete flower before they burst and shed pollen.
- **Eutrophication**: Nutrient enrichment of a given body of water to the point that excessive growth of aquatic plants occurs.
- **Evaporimeter:** A shallow open pan that allows measurement of evaporation. This can then be related to evapo-transpiration from crops or pastures.
- **Evapo-transpiration:** Removal of moisture from soil by evaporation and by transpiration by plants growing in the soil.
- **Exchangeable cations:** Cations that are held by electrical charge to negatively charged clay and humus surfaces. They can be exchanged with other cations in the soil solution.
- **Exchangeable sodium percentage:** Concentration of exchangeable sodium cations expressed as a percentage of the cation exchange capacity.
- **Endemic:** A species restricted to a specified region or locality.
- **F_1& F_2:** When animals / plants of two different breeds or lines are mated, the "First generation of progeny, or first cross, is the F_1 generation. If the F_1 generation males and females re-mated their progeny will be the F_2 generation.
- **Ferrosol:** Soil type with high free iron in the B2 horizon and that lacks strong texture contrast between horizons.
- **Field capacity:** The amount of water remaining in the soil holds after gravitation has drained the free water. Also, the maximum amount of water that a soil can hold under free drainage condition.
- **Field inspection:** An official inspection of field of seed crop conducted by official of certification agency or his authorized agent.
- **Filter strip:** Alter strip is designed to remove soil particles, nutrients and other contaminants as water moves across it.
- **Flocculation:** When a soil is colloidally unstable (i.e., the rate of aggregation is not negligible) then the formation of aggregates is called floculation or coagulation. It is a process wherein colloids come out of suspension is the form of floc or flake, either spontaneosuly or due to the addition of a clarifying agent.
- **Foreign seed:** Undesirable seed which includes weed seeds and other crop seeds.
- **Genes:** Basic units of hereditary material that govern the characteristics of individuals.
- **Genetic diversity:** The variety of genetic information contained in the total genes of individual plants, animals and micro-organisms which inhabit the earth.

- **Genetic engineering:** Manipulating genetic material in such a way as to change the natural course of development to give certain desirable outcomes, or prevent undesirable outcomes.
- **Genetic sterility:** A type of male sterility conditioned by nuclear genes. In contrast to cytoplasmic steriliy, it may be transmitted by either the male or female parent.
- **Genotype:** The genetic makeup of an individual, determined by the genes present.
- **Genus**: A group of related species usually clearly separable from other such groups, or a single species without close relatives; the major taxonomic rank between species and family.
- **Gluten:** Gluten is a general name for the protein found in wheat, spelt, farro, barley triticale etc which is responsible for elasticity in dough and damage body's ability to absorb nutrients from food. Gluten trape CO_2 and expand with it as it is heated.
- **Glycoside:** The term "glycoside" is intended to mean all sugars including glucose found naturally in quinoa including arabinose, glucose, galactose, xylose and glucuronic acid.
- **Glycemic index:** Glycemic index is a measure of how quickly foods raise blood sugar levels. The foods having high glycemic index can stimulate sugar level and contribute obesity.
- **Greenhouse effect:** The effect of the earth's atmosphere on the sun's radiation coming and going from the earth. Most of the sun's rays coming in pass readily through the atmosphere, but the heat waves that are reflected back from the earth are trapped by gases (such as methane, carbon dioxide) in the atmosphere. So the atmosphere acts like the glass in a greenhouse, because it traps the sun's energy.
- **Grow out test:** Test performed to determine the genetic purity of seed as to species or variety.
- **Grubbing:** A manual method to remove weeds. To dig up by the roots.
- **Grass weed:** The leaves are "narrow" as opposed to the "broad" leaves of broad leaved weeds. Also called 'monocots' having one seed leaf opposed to 'dicots' having two seed leaves.
- **Guano:** Decomposed dried excrement of birds and bats. The most commonly known guano comes from islands of the coast of Peru and is derived from the excrement of sea fowl. It is high in nitrogen and phosphate, and at one time was a major fertilizer in this country.
- **Habitat:** The place or type of site in which an organism naturally occurs.
- **Harrowing:** Cultivation with a light tined or disc implement over the soil surface. Used to break down soil clods or remove weeds.
- **Humification:** Conversion of soil organic matter to humus, a stable complex organic compound of varying chemical composition depending on the original residues and the microbial and biochemical processes involved.
- **Humus:** A dark, well–decomposed, stable part of the breakdown of organic matter in the soil.

- **Hybrid:** The offspring of two animals or plants of different races, species, varieties or breeds. They may be fertile or sterile. The greater the difference between the genotypes of the parents, the more likely is sterility. An example is the crossing of a horse and a donkey; the resulting mule is sterile.
- **Hybrid vigour:** Qualities in a hybrid not present in either parent, resulting from the crossing of two dissimilar parents. Increased vigour or other superior qualities arising from the cross breeding of genetically different plants or animals is called hybrid vigour. The process in which hybrid vigour is achieved is called heterosis.
- **Herb:** A plant with a non-woody stem.
- **Herbicide:** A material that kill plants. Herbicides may kill virtually all plants (non-selective) or be quite selective in the way they work. They may be knock down (short-lived) or residual in the soil.
- **Immobilization**: The process of converting inorganic Nitrogen to organic form (typically by microbes which incorporate the N into their own proteins), making the N unavailable to plants.
- **Indicator plants:** A Plant that indicates, in general, and often in a specific manner, a deficiency of plant nutrient in the soil.
- **Inert matter:** Seed like structures from crop and weed plants that are one half the original size or less, badly injured, undeveloped seed like structures of weeds, glumes, stems and other parts, sand dirt and other related substances.
- **Inorganic:** A chemical compound which does not contain carbon chains in its structure, for example, nitrate, phosphate, sulfate and carbonate.
- **Integrated Pest Management (IPM):** A decision support system for crop protection which focuses on long-term prevention or suppression of pest problems with minimum impact on human health, the environment, and non-target organisms. IPM takes into consideration all available pest control techniques and tactics (cultural, mechanical, biological, chemical). IPM emphasizes the growth of healthy crops for better productivity with the least possible disruption to agro-ecosystems.
- **Lot:** A seed lot belonging to variety that is physically identifiable and homogenous. The maximum quantity is specified under Minimum Seed Certification standards.
- **Isolation:** The act of keepping the seed crops away from the sources of physical and genetical contaminant.
- **Land capability class:** A system of classifying land according to its productivity for agricultural use. It gives a land holder a guide to the productive capability of the land and its vulnerability to erosion.
- **LD_{50}:** Measure of toxicity of a chemical. The lethal dose that will kill 50 percent of test animals. The lower the LD_{50}, the more toxic is the chemical.
- **Leaching:** The natural process by which water soluble substances are carried downward through the soil profile into ground water. Leaching accounts for nutrient losses, but can also have the benefit of ridding a soil of excess salts.
- **Line breeding:** Breeding from the same blood-line or within the same family, but not from close relations.

- **Marginal land:** Land that cannot regularly be used for any form of agriculture because it is limited by climatic, soil or slope conditions. This land needs careful management to protect it from degradation.
- **Marginal return:** Extra or additional return from adding an extra unit of produce.
- **Metabolism:** The process occurring inside living cells to generate energy for growth and movement, convert nutrients into cellular structures and undertake other chemical conversion which results in new cells and dispose of cell wastes such as CO_2.
- **Metamorphosis:** Process during pupation by which an immature insect develops into a mature organism through re-arrangement of its structure, such as a caterpillar to a butterfly.
- **Methanogen:** A specialised group of anaerobic bacteria which use hydrogen, organic acids and CO_2 instead of oxygen, to generate energy from decomposition of organic matter. Methanogens create methane as their waste product, a greenhouse gas with 72 times the potential than CO_2 to cause global warming.
- **Mineralization:** Decomposition in soil of organic matter by micro-organisms, which releases the mineral elements as inorganic ions.
- **Minimum tillage:** A soil management system in which weeds are controlled mainly by chemicals instead of cultivation. Conservation farming and reduced tillage involve similar approaches.
- **Myco-herbicide:** A naturally-occurring disease (usually fungi or bacteria) that can be cultivated to produce a more environmentally-safe product than most chemical herbicides.
- **Mycorrhizae:** Fungi forming symbiotic relationships with plant roots, where the organism lives inside the roots enabling access to plant nutrients from sap and its hyphae grow out into soil to absorb nutrients which are transferred to sap. Endo-mycorrhizae have hyphae which lie between the plant cell wall and the inner cell membrane, while ecto-mycorrhizae have hyphae which grow between plant cells, but not against the inner cell membrane, and form a network on the outside of the roots.
- **Micro-nutrients:** Seven nutrients required in very small quantities: Boron, copper, iron, manganese, molybdenum, zinc and chlorine.
- **Mineralization:** The process of converting nitrogen in organic compounds to inorganic form, which makes the Nitrogen available to plants.
- **Nitrification:** A process in which soil micro-organisms form nitrates and nitrites from ammonia or ammonium compounds.
- **Nitrogen fixation:** The transformation of free nitrogen from the air into nitrogenous compounds by various bacteria in the soil. Certain bacteria (*Rhizobium* sp) living in the root nodules of legumes are able to stabilise nitrogen.
- **Nodule:** A small lump on the roots of legumes in which rhizobia bacteria grow and fix nitrogen fromthe air.

- **Nurse crop:** A fast-growing species planted with a slow growing species to encourage it to grow tall and straight or to protect it from grazing or adverse weather.
- **Non-selective herbicide:** A chemical herbicide used for killing / eliminating all types of weed plants (annual and perennial grasses and broad-leaved weeds).
- **Objectionable weed seeds:** Seed, bulblet, tuber of plants designated by Minimum Seed Certification Standards. Weed whose seeds are difficult to seperate once mixed with the crop seed or that genetically contaminate, that which are poisonous or injurious or are having smothering effect on main crop and are difficult to eradicate once established or whose multiplication ratio is so high that they are difficult to control.
- **Off type / rogue:** Plant which does not confirm to varietal characteristics described by the breeder. To designate a plant as a rogue, it is not necessary to identify it as to a variety.
- **Opportunity cropping:** A cropping system which takes advantage of soil moisture buildup to grow more than one crop in a year or to crop in an area that is not normally suited to cropping.
- **Organic:** A chemical compound which contains carbon formed in chains with other nutrients such as N, P, O, H and S, attached to form proteins, sugars, starches, genetic and energy storage molecules. Organic matter can be decomposed by most soil micro-flora to supply energy and nutrients for their growth, reproduction and activities.
- **Organic acid:** An acid which is composed of carbon chains, such as a fatty acid or acetic acid. Generally organic acids do not have pH values as low as inorganic acids such as sulphuric and hydrochloric acids.
- **Organic matter:** Carbon-based materials of either plant or animal origin, which occur in all stages of decomposition in soils.
- **Organic farming:** A farming system based on the use of natural materials for supplying nutrients and protection to plants and animals. Chemicals such as synthetic fertilisers, pesticides or herbicides are not allowed for a farm for organic certificaion.
- **Oxidation:** The combination of oxygen with a substance, or the removal of hydrogen from it.
- **Organic fertilizer**: Strictly speaking, an organic material is one containing carbon. This includes urea and calcium cyanamide which are manufactured synthetically. The term generally applies to products derived from plant or animal sources, such as manure, sewage, slude, castor pomace and process tankage.
- **Perennial weed:** Weeds that return year after year. Some die back in the winter but their roots remain alive and shoots re-appear in spring; some dont die back and grow in size and stature the next season.
- **Perma-culture:** A term for 'permanent agriculture' coined to describe a system of planting, growing and harvesting in harmony with nature. Perma-culture designs use land, plants and animals in the diverse and balanced ways in which they would occur in natural systems.

- **Permanent wilting point:** The amount of water still held in the soil when plants wilt to the extent that they won't recover even if water is added. Soils differ in the amount of water still held at permanent wilting point.
- **pH buffering capacity:** The ability of a soil to absorb acid or lime with little or no change to pH.
- **pH:** A term used to indicate the degree of acidity or alkalinity. Technically, pH is the common logarithm of the reciprocal of the hydrogen ion concentration of a solution. A pH of 7 indicates precise neutrality; higher values indicate increasing alkalinity and lower values indicate increasing acidity.
- **Pheromone:** A sex-attractant hormone or scent produced by animals or insects.
- **Photoperiod:** Number of day-light hours in a day.
- **Photosynthate:** The compounds such as sugars and starches synthesised by plants from carbon dioxide gas using energy from light.
- **Photosynthesis:** The process of plants converting carbon dioxide into carbohydrates by chlorophyll using the energy of sunlight.
- **Phototrophic:** It is the nature of plants which means attracted towards light.
- **Pollination:** The transfer of pollen from the anther to the stigma either within the same flower or between different flowers.
- **Pore space:** The openings in a soil not filled with solid particles. Pore spaces are filled with varying proportions of air and water.
- **Porosity:** The ability of a soil to allow water to move through it. This depends on the quantity and nature of the pore spaces.
- **Primary tillage:** The first working of the soil by implements for cropping or pasture. Also refers to secondary tillage.
- **Post-emergence herbicide**: A herbicide applied after the crop has emerged.
- **Pre-emergence herbicide:** A herbicide applied to the soil after the crop has been sown, but before it emerges.
- **Pre-sowing herbicide**: A herbicide applied to the soil to kill growing or germinating seeds before sowing the crop.
- **Primary nutrients**: The primary nutrients are nitrogen, phosphorus and potassium. They are called primary due to primary importance of manufactures of fertilizers and primary preferance given by the plants relatively in large quantities for healthy growth. They are the most common constituents of commercial fertilizer.
- **Quinoa bran:** The term "quinoa bran" is meant the bran obtained in a commercial mill used to de-bran quinoa for human consumption.
- **Relative humidity:** The amount of water vapour in the air, shown as a percentage of the maximum water vapour that the air could hold at the same temperature and pressure.
- **Rhizobium:** A bacteria that can infect legumes in a beneficial way. The bacteria feed on the plant, but also take nitrogen from the air and change it into a form that the plant can use. Thus legumes have high protein content.

- **Rod weeder:** A light, secondary tillage implement (usually attached behind another implement) that consists of a long rod that revolves just below the soil surface, loosening or flicking weeds out of the soil.
- **Root exudates:** Water soluble compounds, such as amino acids, sugars, vitamins, organic acids, carbohydrates and proteins that roots release in the rhizosphere. Exudates act as messengers, supply food for soil organisms and influence soil moisture in the rhizosphere.
- **Saline soil:** A soil containing enough soluble salts to harm its productivity for plants.
- **Salinity:** An accumulation of salts, usually sodium chloride in soil and water to levels that impact on human and natural assets, such as plants, animals, aquatic ecosystems, water supplies agriculture or infrastructure.
- **Salting:** The accumulation of harmful salts in the soil. There are two types: dryland salinity and irrigation salinity. In both cases, the water table comes close to the surface, the water evaporates and the salts remain. Dryland salinity may be natural, but more usually, like irrigation salinity, it is caused by land management practices. Over clearing of native vegetation can cause dryland salinity.
- **Saponin:** Saponin is meant a compound consisting of a triterpenoid of oleanane structure and one or more glycosides, the glycosides being bound to the triterpenoid at the 3 position and/or at the 28 position.
- **Sapogenin:** The term "sapogenin" is meant the triterpenoid alone without glycosides attached at either the 3 or the 28 position.
- **Saturation:** When the soil root zone has all the pore spaces filled with water, it is saturated or waterlogged.
- **Saturation point:** Soil condition where all the large pores in the soil are filled with water.
- **Secondary tillage:** Tillage operations carried out after primary tillage. It is used for weed control and to prepare the seed-beds for sowing.
- **Seed vigour:** Sum total of those properties of the seed which determines the level of activity and performance of the seed or seed lot during germination and seedling emergence.
- **Seepage area:** Where the water table is close enough to the surface to evaporate and allow salts to accumulate.
- **Selective herbicide:** A herbicide that kills only certain groups of plants. For example, 2,4-D kills broad-leaf plants, but not grasses.
- **Self-pollination:** Pollen is transferred from an anther to a stigma of the same flower.
- **Senescence:** Changes in plants caused by unfavourable growing conditions leading to a halt in growth, and nutrients moving out of the leaves. Growth may begin again if favourable conditions return.
- **Slow-release fertilizer:** A fertiliser that is coated with a fairly insoluble material to provide a slow, predictable release of nutrients.

- **Sodic soil:** Soil with a relatively high proportion of exchangeable sodium. This can cause soil structural problems.
- **Sodicity:** The proportion of exchange sites in a soil or soil layer occupied by sodium ions, expressed as the exchangeable sodium percentage (ESP). Soil with an ESP exceeding 5 is referred to as being sodic and tends to disperse in water.
- **Soil aeration:** A process of encouraging more air to enter the soil, usually by using some type of implement. Aeration is associated with the number and size of pore spaces in the soil.
- **Soil conditioner:** A material is used to improve the chemical or physical quality of the soil. Lime and gypsum are common soil conditioners.
- **Soil degradation:** Soil in which the structure has been damaged, compaction or erosion has occurred. It may also refer to soil acidity and salinity and the loss of nutrients from a soil.
- **Soil erodibility:** The likelihood of a soil to erode.
- **Soil organic carbon:** The carbon fraction associated with soil organic matter, soil microbial biomass, humus and charcoal. Soil organic carbon excludes undecayed plant and animal residues greater than 2 mm in size and inorganic forms, such as lime or carbonates.
- **Soil plasticity:** The capacity of soil to be deformed moulded, or shaped, but not ruptured. The shape of the deformed soil remains after the soil is dried.
- **Soil profile:** Description of each of the layers (soil horizons) in the soil. Examination of the profile is made to determine depth, drainage, texture, structure and other attributes that classify it as soil.
- **Soil productivity:** Productivity is the present capacity of a soil to produce crop yield under a defined set of a management practices. It is measured in terms of the yield in relation to the input of production factors.
- **Soil structure:** The size, shape and arrangement or groupings of soil aggregates and voids. Structural form refers to the 'soil architecture'; structural stability is the ability of a soil to retain its form when immersed in water or compacted, and 'structural resilience' is the ability of a soil to regain structural form by natural processes, such as shrinking and swelling.
- **Soil texture:** A measure of the proportions of different sized soil particles. The relative amounts of sand, silt and clay in a soil.
- **Spray-topping:** The use of herbicides to control the seed set of plants to discourage their reproduction in a pasture or crop.
- **Straight fertilizers:** The fertilizers containing only one plant nutrient e.g., Ammonium sulphate, superphosphate, muriate of potash etc.
- **Strip cropping:** A conservation farming method of growing crops in a series of strips at right angles to the natural water flow across the land, or on the contour. Each strip is of a width designed to cater for cropping machinery, and each strip is treated differently to the strip beside it. A strip may be under fallow, summer crop or winter crop.

- **Stubble mulching:** A conservation farming method in which stubble is kept on the soil surface to protect the soil, but is gradually broken down using appropriate machinery.
- **Symbiosis:** The act of two different organisms living together for mutual benefit. An example is Rhizobium bacteria living within leguminous plants and fixing nitrogen from air into plant available nitrogen, which is made available to host via sap in exchange for nutrients such as sugar and minerals.
- **Salt-Index**: An index used to compare solubilities of chemical compounds. Most nitrogen and potash compounds have a high index, and phosphate compounds have a low index. When applied too close to seed or on foliage, the ones with high indexes can cause salt injury.
- **Secondary Nutrients**: Calcium, magnesium and sulfur.
- **Seed vigour index:** A product of germination percent multiplied by seedling length of a germinated sample.
- **Slurry:** A fluid that contains undissolved solids. Continuous agitation (shaking, stirring, or other physical mixing) is required to maintain uniform concentration throughout the solution volume (without agitation, the solids begin to settle).
- **Soil-amendment:** Any substance such as lime, sulphur, gypsum or sawdust used to improve the productive properties of a soil. Fertilizers are one type of soil amendment. However, many soil amendments such as soil conditioners do not have significant fertilizer value.
- **Selective herbicide:** A chemical product used for eliminating certain types of weeds only (i.e., either grasses or broad-leaved weeds) and not affecting specified crops.
- **Threatened species:** Any species, populations or ecological community's specified in Schedules 1 and 2 of the Threatened *Species Conservation Act, 1995.* Plants and animals may be declared threatened' if their habitat, populations or ecological communities are threatened by outside processes such as clearing.
- **Trace element:** Essential plant nutrients, such as iron, copper, zinc, manganese, boron, molybdenum, chlorine that are required in only very small quantities compared with the major elements.
- **Trickle irrigation:** The application of small quantities of water directly to the root zone through various types of nozzles or drippers.
- **Triticale:** A hybrid winter growing cereal grain plant derived from wheat and rye.
- **Urease:** An enzyme synthesized by some soil bacteria and some plants which splits urea into carbon dioxide and ammonium molecules and makes the nitrogen available for nitrification to nitrate or direct uptake by plants.
- **Vermi-cast:** Nutrient rich excreta from earthworms. It contains chewed and digested decayed organic matter and is valued as a plant food.
- **Vermi-culture:** The process of rearing, breeding and management of earthworms.
- **Vertosol:** Soil types that shrink and crack on drying, and swell on rehydration due to clay minerals.

- **Vesicular-arbuscular mycorrhizae:** Beneficial fungi that develop on many trees, crops and other plants and contribute to nutrient uptake by infecting the plant roots and extending into the soil. Now known more simply as arbuscular mycorrhizae.
- **Virgin soil:** A soil that has not been disturbed significantly or been cultivated.
- **Viticulture:** The cultivation of grapes.
- **Volatilization**: The conversion of a chemical from a liquid or solid to a gas or vapour. By this process, N fertilizers such as ammonia or urea are lost to the atmosphere under certain conditions.
- **Volunteer plants:** A plant of the crop species same as that of crop which comes up on account of being self-grown from previous crop.
- **Water-holding capacity:** The ability of soil to hold water. The moisture-holding capacity of sandy soils is usually low,while that of clay soils is high.
- **Wilting point:** Soil moisture content at which a plant starts to wilt. This varies with soil texture.
- **Weed mutation:** A permanent structural alteration in the DNA or RNA of a weed that affects its entire off-spring via seeds.
- **Weed shifts:** A change in the weed community within a field i.e., relative abundance or type of weeds. This can be the result of a management practice like herbicide use or any other phenomenon that brings about a change in weed species composition. Species or biotypes adapted to current weed management practices increase, whereas weeds susceptible to those practices decrease.
- **Yeast:** A microscopic, single-celled fungus which can grow with or without oxygen. Yeasts can ferment sugars to produce acid and alcohol if growing without oxygen.
- **Zero-till:** Sowing a crop without any primary tillage. Chemicals, grazing or both are used to reduce weed growth, and seed is sown into an unprepared seedbed.

SELECTED REFERENCES

1. Abugoch, J. L. E., Quinoa (*Chenopodium quinoa* Willd.), Composition, Chemistry, Nutritional, and Functional Properties. *Advances in Food and Nutrition Research*, Vol. 58, pp. 2009, 1–31.
2. Ahamed, T., Singhal, R., Kulkarni, P., and Pal, M. (1998), A lesser known grain, *Chenopodium quinoa*: A review of the chemical composition of its edible parts. *Food Nutrition Bulletin*, Vol.19, pp. 61–70.
3. Alvarez-Jubete, L., Arendt, E.K., Gallagher, E (2009), Nutritive value and chemical composition of pseudocereals as gluten-free ingredients. International *Journal of Food Sciences and Nutrition*, Vol. 60, pp. 240–257.
4. Alvarez-Jubete, L., Arendt, E.K., Gallagher, E (2010), Nutritive value of pseudocereals and their increasing use as functional glúten-free ingredients. *Trends in Food Science and Technology*, Vol. 21, pp. 106–113.
5. Alvarez-Jubete, L., Auty, M., Arendt, E.K., Gallagher, E (2010), Baking properties and microstructure of pseudocereal flours in gluten-free formulations. *European Food Research and Technology*, Vol. 230, pp. 437–445.
6. Alvarez-Jubete, L., Wijngaard, H., Arendt, E.K., Gallagher, E (2010 b), Polyphenol composition and *in vitro* antioxidant activity of amaranth, quinoa, buckwheat and wheat as affected by sprouting and baking. *Food Chemistry*, Vol. 119, pp. 770–778.
7. Ando, H., Chen, Y., Tang, H., Shimizu, M., Watanabe, K., and Miysunaga, T (2002), Food components in fractions of quinoa seed. Food Science Technology Research, Vol. 8, pp. 80–84.
8. Atwell, W.A., Patrick, B.M., Johnson, L.A., and Glass, R.W (1983), Characterization of quinoa starch. *Cereal Chemistry*, Vol. 60, pp. 9–11, 1983.
9. Arun K.S (2007), Bio-fertilizers for sustainable agriculture. Mechanism of P-solubilization Sixth edition, *Agribios publishers, Jodhpur, India*, 196–197.
10. Asada K (1999), The water-water cycle in chloroplasts. Scavenging of active oxygens and dissipation of excess photons. *Annu Rev Plant Physiol Plant Mol Biol* 50 : 601–639.
11. Barcelo J, and Poschenrieder C (2004). Structural and ultrastructural changes in heavy metal exposed plants. In: Prasad MNV (ed): Heavy metal stress in plants, 3rd Edn. Springer, Berlin, pp. 223–248.
12. Baker, L.A., and Rayas-Duarte, P (1998). Freeze-thaw stability of amaranth starch and the effects of salt and sugars. *Cereal Chemistry*, Vol.75, pp. 301–307.

13. Becker, R., Wheeler, E. L., Lorenz, K., Stafford, A.E., Grosjean, O.K., Betschart, A.A. and Saunders, R.M (1981). A compositional study of amaranth grain. *Journal Food Science*, Vol. 46, pp. 1175–1180.

14. Berghofer E, and Schoenlechner R. Grain Amaranth. In: Belton P, Taylor J (2002): Pseudo-cereals and less common cereals, grain properties and utilization potential. Springer - Verlag, pp. 219–60.

15. Berti, C., Riso, P., Monti, L., and Porrini, M (2003): *In vitro* starch digestibility and *in vitro* glucose response of gluten-free foods and their gluten counterparts. *European Journal of Nutrition*, Vol. 43, No.4,, pp. 198–204.

16. Bressani, R. Amaranth. In B. Caballero (2003 ed.): Encyclopedia of food sciences and nutrition. Oxford: Academic Press, pp. 166–173.

17. Brinegar, C., Sine, B., and Nwokocha, L (1996): High-cysteine 2S seed storage proteins from quinoa (*Chenopodium quinoa*). Journal of Agricultural and Food Chemistry, Vol. 44, No.7, pp. 1621–1623.

18. Bruni, R., Medici, A., Guerrini, A., Scalia, S., Poli, F., Muzzoli, M., and Sacchetti, G (2001): Wild *Amaranthus caudatus* seed oil, a nutraceutical resource from Ecuadorian flora. Journal of Agricultural and Food Chemistry, Vol. 49, pp. 5455–5460.

19. Berry, W.L, and Wallace A (1981), Toxicity - the concept and relationship to the dose–response curve. *J. Plant Nutr* 3:13–19.

20. Bertrand M, and Poirier I (2005), Photosynthetic organisms and excess of metals. Photosynth. 43, pp. 345–353.

21. Caperuto, L., Amaya-Farfan, J., and Camargo, C (2000), Performance of quinoa (*Chenopodium quinoa* Willd.) flour in the manufacture of gluten-free spaghetti. *Journal of the Science of Food and Agriculture*, Vol. 81, pp. 95–101.

22. Chauhan, G.S., Eskin, N.A.M., and Tkachuk, R (1992), Nutrients and antinutrients in quinoa seed. Cereal Chemistry, Vol. 69, pp. 85–88.

23. Chauhan, G.S., Zillman, R.R., and Eskin, N.A.M (1992), Dough mixing and breadmaking properties of quinoa-wheat flour blends. International *Journal of Food Science and Technology*, Vol. 27, pp. 701–705.

24. Choi, H., Kim, W., and Shin, M (2004), Properties of Korean amaranth starch compared to waxy millet and waxy sorghum starches. *Starch/Starke*, Vol. 56, pp. 469–477.

25. Copeland, L (2009), Structural characterization of wheat starch granules differing in amylase content and functional characteristics. *Carbohydrate Polymers*, Vol. 75, pp. 705–711.

26. Coulter, L.A., Lorenz, K (1991), Extruded corn grits-quinoa blends: I Proximate composition, nutritional properties and sensory evaluation. *Journal of Food Processing and Preservation*, Vol. 15, pp. 231–242.

27. Clemens S (2006), Toxic metal accumulation, responses to exposure and mechanisms of tolerance in plants. *Biochimie* 88, pp. 1707–1719.

28. Danz, R.A. and Lupton, J.R (1992), Physiological effects of dietary amaranth (*Amaranthus cruentus*) on rats. *Cereal Food World*, Vol. 37, pp. 489 – 494.

29. Dodok, L., Modhir, A.A., Buchtová, V., Halásová, G. and Poláček, I (1997), Importance and utilization of amaranth in food industry. Part 2. Composition of amino acids and fatty acids. *Nahrung*, 41, pp. 108–110.

30. Duffus JH (2002), "Heavy metal" – a meaningless term? *Pure Appl Chem* 74, pp. 793 – 807.

31. Early, D.K (1990), Amaranth production in Mexico and Peru. In Janick and J.E. Simon (eds), Advances in new crops. Timber Press, Portland, OR, pp. 140 – 142.

32. Ernst WHO (2004), The use of higher plants as bioindicators. In: Markert BA, Breure AM, Zechmeister GH (eds) Bioindicators and biomonitors. Elsevier Science, Amsterdam, pp. 423–464.

33. Fodor F (2002), Physiological responses of vascular plants to heavy metals. In: Prasad MNV, Strzalka K (Eds) Physiology and biochemistry of metal toxicity and tolerance in plants. Kluwer Academic Publisher, Dortrech, pp. 149–177

34. Gallardo, M., Gonzales, A., Ponessa and G. Morfologia del fruto y semilla de *Chenopodium quinoa* Willd. (Quinoa). Chenopodiacea. Lilloa 39, 1, 1997.

35. Gamel, T. H., Linssen, J. P., Alink, G. M., Mossallem, A. S., and Shekib, L. A (2004), Nutritional study of raw and popped seed proteins of *Amaranthus caudatus* and *Amaranthus cruentus*. *Journal Science and Food Agriculture*, Vol. 84, pp. 1153–1158.

36. Hagemeyer J (2004), Ecophysiology of plant growth under heavy metal stress. In: Prasad MNV (ed): Heavy metal stress in plants, 3rd Edn. Springer, Berlin, pp. 201–222.

37. Hozová, B., Buchtová, V., Dodok, L., and Zemanovic, J (1997), Microbiological, nutritional and sensory aspects of stored amaranth biscuits and amaranth crackers. Nahrung, 41, pp. 151 – 158.

38. Housecroft CE, and Sharpe AG (2008): Inorganic Chemistry. Prentice Hall, Harlow Kovacs E, Nyitrai P, Czovek P, Ovari M, Keresztes A (2009): Investigation into the mechanism of stimulation by low-concentration stressors in barley seedlings. *J Plant Physiol*, 166 : 72–79.

39. Irving, W.D., Betschart, A.A. and Saunders, R.M (1981): Morphological studies on *Amaranthus cruentus*. Journal of Food Science, 46, (4), pp. 1170 – 1174.

40. Jahaniaval, F., Kakuda, Y. and Marcone, M.F (2000), Fatty acid and triacylglycerol composition of seed oils of Five Amaranthus accessions and their comparison to other oils. *JAOCS*, 77 (8), pp. 847 – 852.

41. Kong, X., Bao, J., and Corke, H (2009): Physycal properties of Amaranthus starch. *Food Chemistry*, 113, pp. 371 – 376.

42. Koziol, M. J (1992), Chemical composition and nutritional evaluation of quinoa (*Chenopodium quinoa* Willd). *Journal of Food Composition Analysis*, 5, pp. 35 – 68.

43. Koziol, M.P (1993), Quinoa: A potential new oil crop. In: "New crops". J. Janick and J.E. Simon, (eds.), Wiley, New York, pp. 328 - 36.

44. Leon-Camacho, M., Garcia-Gonzalez, D. L., and Aparicio, R (2001), A detailed and comprehensive study of amaranth (Amaranthus cruentus L.) oil fatty profile. *Eur. Food Research and Technology*, Vol. 213, pp. 349 – 355.

45. Lindeboom, N., Chang, P.R., Falk, K.C., and Tyler, R.T (2005): Characteristics of starch from eight quinoa lines. *Cereal Chemistry*, Vol. 82, pp. 216 – 222.

46. Lorenz, K (1981), Amaranthus hypochondriacus – Characteristics of the starch and baking potential of the flour. *Starch/Starke*, Vol. 33, pp. 149 – 153.

47. Lorenz, K., and Coulter, L (1991), Quinoa flour in baked products. *Plan Foods for Human Nutrition*, Vol. 41, pp. 213–223.

48. Lee KC, Cunningham BA, Paulson GM, and Liang GH, Moore RB (1976), Effects of cadmium on respiration rates and activities of several enzymes in soybean seedlings. *Plant Physiol* 36, pp. 4 – 6

49. Lösch R (2004), Plant mitochondrial respiration under the influence of heavy metals. In: Prasad MNV (ed): Heavy metal stress in plants, 3rd Edn. Springer, Berlin, pp. 182–200.

50. Marschner H (1995), Mineral nutrition of higher plants. Oxford University Press, London.

51. Marcone, M.F (1999), Evidence confirming the existence of a 7S globulin-like storage protein in *Amaranthus Hypocondriacus* seed. *Food Chemistry*, 65, pp. 533 – 542.

52. Marcone, M.F., and Rickey, Y.Y, Evidence for the phosophorylation and glycosylation of the amaranth 11S globulin (Amaranthin). Journal of Food Biochemistry, 21, pp. 341 – 369.

53. Moreau, R., Whitaker, B., and Hick, K (2002): Phytosterols, phytostanols, and their conjugates in foods: Structural diversity, quantitative analysis, and health-promoting uses. *Progress in Lipid Research*, 41 (6), pp. 457 – 500.

54. John Wiley, Chichefster Mysliwa-Kurdziel B, Prasad, MNV and Stralka K (2004), Photosynthesis in heavy metal stress plants. In: Prasad MNV (ed) Heavy metal stress in plants, 3rd Edn. Springer, Berlin, pp. 146–181.

55. Naumann B, Eberius M and Appenroth, K-J (2007), Growth rate based dose–response relationshipsand EC-values of ten heavy metals using the duckweed growth inhibition test (ISO 20079) with *Lemna minor* L. clone St. J *Plant Physiol* 164 : 165 6–1664.

56. Nieboer E and Richardson DHS (1980), The replacement of the nondescript term "heavy metals" by a biologically and chemically significant classification of metal ions. *Environ Poll Series B - Chem Phys* 1:3–26.

57. Nyitrai P, Mayer M, Ovari M, Keresztes A (2007), Involvement of the phosphoinositide signaling pathway in the anti-senescence effect of low-concentration stressors on detached barley leaves. *Plant Biol* 9 : 420 – 426.

58. Park, S.H., Maeda, T. and Morita, N (2005), Effect of whole quinoa flours and lipase on the chemical, rheological and breadmaking characteristics of wheat flour. Journal of Applied Glycoscience, Vol. 52, pp. 337 – 343.

59. Pedersen, B., Kalinowski, L.S. and Eggum, B.O (1987): The nutritive value of amaranth grain (*Amaranthus caudatus*) 1. Protein and minerals of raw and processed grain. *Plant Foods for Human Nutrition*. Vol. 36, pp. 309 – 324.

60. Prego, I., Maldonado, S. and Otegui, M (1998): Seed structure and localization of reserves in *Chenopodium quinoa*. *Annals of Botany*, Vol. 82, pp. 481 – 488.

61. Pearson RG (1968), Hard and soft acids HSAB: Fundamental principles. *J Chem Educ* 45 : 581–587

62. Poschenrieder C and Barcelo J (2004), Water relations in heavy metal stressed plants. In: Prasad MNV (ed) Heavy metal stress in plants, 3rd Edn. Springer, Berlin, pp. 249–270.

63. Qian, J.Y. and Kuhn, M (1999), Characterization of *Amaranthus cruentus* and *Chenopodium quinoa* starch. Starch/Starke, Vol. 51, No. 4, pp. 116–120.

64. Qureshi, A.A., Lehman, J.W. and Peterson, D.M (1996), Amaranth and its oil inhibit cholesterol biosynthesis in 6-week-old female chickens. *Journal of Nutrition*, Vol. 126, No. 8, pp. 1972–1978.

65. Ranhotra, G.S., Gelroth, J.A., Glaser, B.K., Lorenz, K.J. and Johnson, D.L (1993), Composition and protein nutritional quality of quinoa. *Cereal Chemistry*, Vol. 70, No, 3, pp. 303 – 305.

66. Rosell, C.M., Cortez, G. and Repo-Carrasco, R (2009), Bread making use of the Andean crops quinoa (*Chenopodium quinoa*), kañiwa (*Chenopodium pallidicaule*), kiwicha (*Amaranthus caudatus*), and tarwi (*Lupinus mutabilis*). *Cereal Chemeistry*, Vol. 86, No. 44, pp. 386 – 392.

67. Ruales, J. and Nair, B.M (1992), Nutritional quality of the protein in quinoa (*Chenopodium quinoa*, Willd.) seeds. Plant Foods for *Human Nutrition*, Vol. 42, pp. 1 – 11.

68. Ruales, J. and Nair, B.M (1993), Content of fat, vitamins and minerals in quinoa (*Chenopodium quinoa*, Willd.). *Food Chemistry*, Vol. 48, pp. 131–136.

69. Ruales, J. and Nair, B.M (1994), Properties of starch and dietary fibre in raw and processed quinoa (*Chenopodium quinoa, Willd.*) seeds. *Plant Foods for Human Nutrition*, Vol. 45, pp. 223–246.

70. Ryan, E., Galvin, K., O´Connor, T., Maguire, A. and O´Brien, N (2007), Phytosterols, squalene, tocopherol content and fatty acid profile of selected seedds, grains, and legumes. *Plant Food Human and Nutrition*, Vol. 62, pp. 85–91.

71. Rengel Z (2004), Heavy metals as essential nutrients. In: Prasad MNV (ed) Heavy metal stress in plants, 3rd Edn. Springer, Berlin, pp. 271–294.

72. Salas, A.G.V (2011), Development of quinoa (*Chenopodium quinoa* Willd.) based products with functional characteristic. 106 p. Master Science Thesis – Pharmaceutical Sciences School, University of Sao Paulo, Sao Paulo-Brazil (*Advisor: Suzana Caetano da Silva Lannes*).

73. Sauer, J.D (1967), The grain amaranth and their relatives: A revised taxonomic and geographic survey. *Ann. Missouri Bot. Gard*, Vol. 54, pp. 103–37.

74. Schoenlechner, R., Drausinger, J., Ottenschlaeger, V., Jurackova, K and Berghofer, E (2010): Functional properties of gluten-free pasta produced from amaranth, quinoa and buckwheat. *Plant Foods for Human Nutrition*, Vol. 65, pp. 339–349.

75. Schoenlechner, R., Siebenhandl, S. and Berghofer, E. Pseudocereals. In: Arendt, E.K., Bello, F.D (2008): Gluten-free cereal products and beverages. Academic Press. Chapter 7. pp. 149–190.

76. Shin, D.H., Heo, H.J., Lee, Y.J. and Kim, H.K (2004), Amaranth squalene reduces serum and liver lipid levels in rats fed a cholesterol diet. *Brazilian Journal of Biomedical Sciences*, Vol. 61, No. 1, pp. 11–14.

77. Sindhuja, A., Sudha, M.L. and Rahim, A (2005), Effect of incorporation of amaranth flour on the quality of cookies. *European Food Research and Technology*, Vol. 221, pp. 597–601.

78. Simopoulos, A. P (2001), Evolutionary aspects of diet, the omega-6/omega-3 ratio and genetic variation: nutritional implications for chronic diseases. *Biomedicine and Pharmacotherapy*, Vol. 60, pp. 502–507.

79. Spehar, C.R. and Souza, P.I.M (1993), Adaptação da quinoa (*Chenopodium quinoa* Willd.) ao cultivo nos cerrados do Planalto Central: resultados preliminares. *Pesquisa Agropecuária Brasileira*, Vol. 28, pp. 635–639.

80. Spehar, C.R., Teixeira, D.L., Cabezas, W., Erasmo and E.A. Amaranto BRS Alegria (2003), alternativa para diversificar os sistemas de produção. *Pesquisa Agropecuária Brasileira*, Vol. 38, pp. 659–663.

81. Schützendübel A, and Polle A (2002), Plant responses to abiotic stress: heavy metal-induced oxidative stress and protection by mycorrhization. J. Exp. Bot. 53: 1351–1365.

82. Sharma RJ, and Agrawal M (2005), Biological effects of heavy metals: An overview. *J Exp Bot*. 26 (2 suppl): 301–313.

83. Shaw BP, Sahu SK and Mishra RK (2004), In: Prasad MNV (ed): Heavy metal stress in plants 2nd Edn. Springer, Berlin, pp. 84–126.

84. Subba Roa N.S (2001), An appraisal of biofertilizers in India. The biotechnology of biofertilizers, (ed.) S.Kannaiyan, *Narosa Pub. House.*

85. Taylor, J. R. N., Parker and M. L. Quinoa 2002. In P. S. Belton, and J. R. N. Taylor (eds.), Pseudocereals and less common cereals: Grain properties and utilization. Berlin: Springer Verlag, pp. 93–122.

86. Tosi, E.A.; Ciappini, M.C. and Mascarelli, R (1996), Utilización de la harina integral de amaranto (*Amaranthus Cruentus*) In: La Fabricación de galletas para celíacos. Alimentaria, pp. 49–55.

87. Tosi, E.A., Ré, E.D., Masciarelli, R., Sánchez, H., Osella, C., de La Torre, M.A (2002): Flour tested as wheat flour supplementation in mold breads. *Lebensmittel-Wissenschaft und Technologie*, Vol. 35, pp. 472–475.

88. Venkatashwarlu B (2008), Role of bio-fertilizers in organic farming. Organic farming in rainfed agriculture: Central institute for dry land agriculture, Hyderabad, 85–95.

89. Van Assche F and Clijsters H (1990), Effects of heavy metals on enzyme activity in plants. *Plant cell Environ* 13 : 195–206

90. Yawalkar, K.S., Agarwal, J.P. and Bokde, S (1972), Manures and fertilizers (1972 edition).

91. Wani S.P. and Lee K.K (1995), Micro-organisms as biological inputs for sustainable agriculture in Organic Agriculture (Thampan, P.K.ed.). Peekay Tree Crops Development Foundation, Cochin, India, 39–76.

92. Wani, S.P. and Lee K.K (2002), Population dynamics of nitrogen fixing bacteria associated with pearl millet (*P. americanum* L.), in biotechnology of nitrogen fixation in the tropics, University of Pertanian, Malaysia, 21–30.

ANNEXURE - I

Comparative nutritional quality of quinoa and other traditional food grains

Crop	Water	Crude Protein	Fat	Carbohydrate	Fibre	Ash
	%					
Quinoa	12.6	13.8	5.0	59.7	4.1	3.4
Barley	9.0	14.7	1.1	67.8	2.0	5.5
Buckwheat	10.7	18.5	4.9	43.5	18.2	4.2
Corn	13.5	8.7	3.9	70.9	1.7	1.2
Millet (Pearl)	11.0	11.9	4.0	68.6	2.0	2.0
Oat	13.5	11.1	4.6	57.6	0.3	2.9
Rice	11.0	7.3	0.4	80.4	0.4	0.5
Rye	13.5	11.5	1.2	69.6	2.6	1.5
Wheat	10.9	13.0	1.6	70.0	2.7	1.8

ANNEXURE - II

Comparison of Amino acids' composition of quinoa grains with barley, soybean and wheat

Amino acid	QUINOA	BARLEY	SOYBEAN	WHEAT
	Mg/g protein			
Arginine	77.3	50.1	69.5	–
Aspartic acid	80.3	62.5	136.3	94
Cystine	14.4	22.1	12.1	20.5
Glycine	49.2	36.2	38.6	45.5
Glutamic acid	132.1	261.2	151	195.1
Histidine	28.8	–	26.7	23.5
Isoleucine	35.7	–	44.5	43.2
Leucine	59.5	–	72	82.8
Lysine	54.2	37.2	57.8	36.2
Methionine	21.8	19.2	10.6	23.5
Phenylalanine	42	56.1	49.2	53.5
Serine	40.2	42.2	50	52.6
Threonine	29.8	34	38.6	35.8
Tryptophan	11.8	16.6	12.2	11.5
Tyrosine	18.9	28.7	36.2	33.4
Valine	42.1	49	47.6	61.1
Alanine	41.6	39	42.2	58

SANNEXURE - III

Pattern of essential amino acids in quinoa protein as compared with wheat, soybean and skimed milk

Amino Acid	Amino acid content (g/100g protein)			
	Quinoa	Wheat	Soybean	Skim Milk
	---------------------- % ----------------------			
Isoleucine	4.0	3.8	4.7	5.6
Leucine	6.8	6.6	7.0	9.8
Lysine	5.1	2.5	6.3	8.2
Phenylalanine	4.6	4.5	4.6	4.8
Tyrosine	3.8	3.0	3.6	5.0
Cystine	2.4	2.2	1.4	0.9
Methionine	2.2	1.7	1.4	2.6
Threonine	3.7	2.9	3.9	4.6
Tryptophan	1.2	1.3	1.2	1.3
Valine	4.8	4.7	4.9	6.9

ANNEXURE - IV

Mineral contents in Quinoa grains as compared to traditional food grains

Crop	Ca	P	Mg	K	Na	Fe	Cu	Mn	Zn
	----------%---------				-----------PPM----------				
Quinoa	0.19	0.47	0.26	0.87	115	205	67	128	50
Barley	0.08	0.42	0.12	0.56	200	50	8	16	15
Corn	0.07	0.36	0.14	0.39	900	21	–	–	–
Wheat	0.05	0.30	0.16	0.52	900	50	7	–	14

ANNEXURE - V

Role of vitamins in human body

S.No.	Name of vitamin	Nature	Functions
Fat soluble, Stored in fatty tissues & liver upto 6 months, transported in areas of requirement within body			
1	Vitamin-A (Beta-carotene)	Belongs to pigment-Carotenoids associated with arthritis, heart and progression of cancer	Helps to form and maintain healthy teeth, bones, soft tissues, mucus membrane and skin, boost immune system
2	Vitamin-D (Sunshine vitamin)	10–15 minutes of sunshine three times a week is enough to produce the body's requirement of this vitamin.	Maintain the bones and level of calcium & phosphorus in blood
3	Vitamin-E (Tocopherol)	Anti-oxidant,	Helps in formation of Red Blood Cells and use of vitamin-A
4	Vitamin-K	Not listed in essential vitamins	Helps blood coagulation
Water soluble, not stored in the body, travel in blood stream and left body through urine			
1	Vitamin B1 (Thiamin)	Necessary during pregnancy & breast-feeding	Essential for conversion of carbohydrates into energy, heart functions & healthy nerve cells.
2	Vitamin B2 (Riboflavin)	Works with other B-vitamins	Helps in production of Red Blood Cells & growth
3	Vitamin B3 (Niacin)	–	Maintains healthy skin & nerves, have cholesterol lowering effect at higher doses
4	Pantothenic acid (Vitamin-B5)	–	Helps in metabolism of food, production of hormones & cholesterol
5	Pyridoxine (Vitamin-B6)	–	Helps in formation of blood cells, maintain brain function and play role in protein,
6	Biotin (Vitamin-B7)	–	Helps in metabolism of protein & charbohydrate, Production of hormones cholesterol & fatty acids

7	Folate (Vitamin-B9)	Works with B-12	Helps in form blood cells, formation of DNA, its low level is linked to birth defects.
8	Cyanocobalamin (Vitamin-B12)	Water soluble but stored in fatty tissues	Play part in DNA & cells synthesis, cardiovascular protection and energy production
9	Ascorbic acid (Vitamin-C)	Anti-oxidant	Promotes healthy teeth, gums, wound healing, absorption of iron and maintain healthy tissues.

ANNEXURE - VI

Role of Minerals in human body

S.No.	Name of minerals	Nature	Functions
Macro-minerals			
1	Calcium	Abundant mineral in body	Keeps bones & teeth strong, support skeleton structure, receive neurotransmitters during communication, helps to conduct electricity in body, maintain regular heart beat.
2	Phosphorus	Second most abondant mineral in body after calcium, 85% found in bones & teeth	Gives strength in teeth and bones, repair body cells & tissues, provide structure & strength.
3	Potassium	It is blood mineral called electrolite, found within the cells, kidneys regulate potassium in body	Conducts electricity in body, helps generate muscles cotraction & regulate heart beat, prevent swelling of cells, helps synthesis of protein & important for building muscles, help balance the negative effects of too much sodium.
4	Magnesium		Maintains normal nerves & muscles function, support healthy immune system, keep heartbeat steady, provides strength to bones, regulate blood glucose level & blood pressure, helps production of protein & energy.
5	Sulphur	3rd most abundant mineral in body, key component of 4 amino-acids i.e., methionine, cysteine, cystine & taurine.	Helps in solidifying protein structure, creat protein for cells,

6	Sodium	It is both electrolite & mineral, 85% sodium of body found in blood & lymph fluid, low sodium cause muscle cramps, vomitting, diziness and lead shock & coma.	Helps to keep the water & electrolite balance of body, regulate work of nerves & muscles, maintain body's fluid balance
7	Chloride	It is electrolite in blood	Helps to balance fluid inside & outside cells, maintain pH of body fluid, blood pressure & blood volume.

Micro / trace minerals			
1.	Iron	70% iron present in RBC (hemoglobin) & muscle cells (myoglobin), liver & spleen are iron richest organs, Women require more iron due to its major loss during menstruation	Play vital role in metabolic process, oxygen transport through hemoglobin, essential for blood production.
2.	Iodine	60% iodine stored in thyroid gland	Controls functioning of thyroid gland, regulate utilisation of calories & prevent its storage as excess fat.
3.	Manganese	Concentrated in liver, pancreas, kidneys & bones. Deficiency cause joint pains, poor bone ealth & fertility problem.	Contributes to healthy bone structure, bone metabolism & helps to create enzymes for building bones.
4.	Zinc	Powerful antioxidant, found in cells throughout body	Needed for immune system for proper work & fight off invading bacteria & viruses, involve directly in cell division, cells growh, wound healing & breakdown of carbohyydrates.
5.	Cobalt	Stored in plasma, spleen, pancreas & kidneys	Repairs myelin (which protect nerve cells), helps formation of hemoglobin, Red Blood Cells & vitamin 12 & activate enzymes.
6.	Copper	Present in all tissues, toxic in excess	Aids in iron absorption, keep bones, blood vessels, nerve & immune healthy, require for cellular respiration, biosynesis of neurotransmitter, pigment formation.

ANNEXURE - VII

Role of Fatty acids in human body

S.No.	Name of fatty acids	Nature	Functions
Essential Fatty acids (Not synthesize in body, provide building blocks to body, increases & decreases inflammation)			
1.	Alpha-linolenic (Omega-3)	Poly-unsaturated fat, found in foods, nuts, vegetable oils abudantly, it is part of cell membrane throughout body	Binds receptors in cells & regulate their functions, prevent heart diseases / stroke & rheumatiod arthritis, play role in brain function, protective role in cancer, lower blood pressure, reduces blood triglyceride level & plays role in weight management, important for brain deveopment in babies.
2.	Linoleic (Omega-6)	Poly-unsaturated fat	Plays crucial role in brain function, stimulate skin & hair growth, maintains bone health, regulates metabolism and maintains reproductive system
Non-essential fatty acids (Produced naturally in the body, not needed for vital functions, stored underneath skin)			
3.	Omega-9	Mono-unsaturated & abundant fat in cells of body	Promotes heart health, improves immune system and have metabolic health
4.	Saturated fatty acids	They are un-healthy, raise cholesterol level	Plays key role in immune health
5.	Trans fatty acids	Worst kind of fat	Raises bad LDL, risk of heart disease & lower good HDL

ANNEXURE - VIII

Economics of organic cultivation of Quinoa in India

S.No.	Particulars	Expenditure INR
1	Total cost of production Acre^{-1}	21835/-
2	Average yield of Quinoa Acre^{-1}	1000 Kg
3	Average sale rate of un-processed quinoa grains (Rs. Kg^{-1})	100/-
4	Gross income Acre^{-1} (Rs.) (1000 x 100)	100000/-
5	Net return Acre^{-1} (Rs.)	**78,165/-**

Details of cost of organic cultivation

S.No.	Particulars	Expenditure INR
1.	Field preparation and layout • 2 ploughings @ Rs. 400/- each • Field layout: 4 labour @ Rs. 300/- per day each	2000/-
2.	Seed cost – 1 Kg (@ Rs. 1500/- each Kg)	1500/-
3.	Seed sowing cost 5 women labour Acre^{-1} (@ Rs. 200/- per day each)	1000/-
4.	Organic inputs: • 5 Kg Harit Vardan (MRP - Rs. 1650 per Kg, 20% subsidy) i.e. @ Rs. 1320/- Kg^{-1} • 30 Kg Nano-Compost (@ Rs. 75/- Kg^{-1}, • 1 Liter Bioplantomin (Micro-nutrient, Cost-1950/- Liter $^{-1}$, 20% subsidy) i.e. Rs. 1560/- Kg^{-1}	10410/-
5.	Irrigation – 3 (@ Rs. 400/- per irrigation)	1200/-
6.	Weed control (2 manual weeding) (10 women labour @ Rs. 200/- each day^{-1})	2000/-
7.	Plant protection measures	350/-
8.	Harvesting (5 women labour @ Rs. 200/- each day^{-1})	1000/-
9.	Threshing, cleaning & grading (10 women labour @ Rs. 200/- each day^{-1})	2000/-
10.	Packing (25 bags @ Rs. 15/- per bag)+ other charges	375/-
	Total	**21,835/-**

ANNEXURE - IX

Prime contributors to Quinoa development in India

S.No.	Name	Address
1.	Dr. Dharm Singh Director General	Indian Medicinal Plants Marketing Federation, New Delhi
2.	Shri A.K.Singh, IAS Add. Commissioner	Govt. of Uttar Pradesh, Lucknow (UP)
3.	Dr. S. L. Jadhav Director (Agri / Horti)	Directorate of Agriculture, Government of Maharashtra, Shivaji nagar, Pune
4.	Shri Mukesh Dave Regional Director	Sarvati Devi Memorial Trust, Regional Center, Sangamner (Maharashtra)
5.	Shri Pradeep Dwivedi Managing Director-cum-Food Tehnologist	Nutrelis Foods, Noida (UP)
6.	Shri Dev Singh & Shri Dinesh Singh Directors	3D Industries Pvt. Ltd, Sarvati Devi Research Complex (L-21), New Mahavir Nagar, New Delhi -110018
7.	Shri Mahadevan Pillai Chief Executive Officer	Federal Union of Quinoa growers and Traders of India, Ahmedabad, Gujarat
8.	Shri P.Jeevan Prashanth Co-Founder	Orillet Foods International Shed No. 1, Sagar Estate, BKS Mandal, Vadiampeta – 515731, Ananthapur District, Andhra Pradesh
9.	Shri A.G. Veera Reddy, Managing Director	V V Herbs and Organics, 25/684/150, Padmavothi nagar, Nandyal, Karnool district, Andhra Pradesh - 518501
10.	Shri Heer Singh Purohit	Village – Bawatra, District - Jalor (Rajasthan)

ANNEXURE - X

Establishment of Quinoa Processing Unit

= Model Project Report =

1. **Project title :** Establishment of Quinoa processing unit.
2. Project Implementing Agency (PIA) :

- **Name :** Sarvati Devi Memorial Trust
- **Legal status :** National Voluntary Organization
  (Registered under Societies Registration Act of 1860 as applicable to Delhi).
- **Year of registration :** 19 May, 1988
- **Administrative office :** Sarvati Devi Research Complex, L -21, New Mahavir Nagar, New Delhi-110018

Contact

- **President :** Dr. Dharm Singh, M- 8860980322, 8588986083
- **Regional Director (M.S.) :** Shri Mukesh Dave, M- 8141547637

3. **Name of work to be done :** Processing, packaging & marketing of quinoa grains
4. **Processing capacity of unit**

Quantity to be processed

- **Per day capacity :** 100 tones
- **Per month capacity : 3000** tones
- **Per Year capacity : 36**000 Tons
- **Value of processed grains : INR 1620 Crores** per year (@ Rs. 450/- Per Kg, 450000/- per ton, Lowest price of organic quinoa in India).

5. **Year of factory establishment :** 2017 – 18
6. **Factory location (proposed):** Mathura (UP) / Mumbai (Maharashtra)

7. Introduction:

India is a self-sufficient and largest producer of food grains in the world. Due impending climatic variability, drought, flood, water logging, soil salinity, nutrient deficiency and climate change often resulted failure / uncertainty in production and downfall in traditional food grain crops leading to its shortage and poor availability of food to the people. In the situation of fast growing population in India, food security is a serious concern and major challenge as over 195 million people presently not getting adequate two square meals in a day. Quinoa is highly resilient to agro-ecological extremes (soils, rainfall, temperature, altitude), tolerant to frost, drought and salinity and rich in protein, tocopherols, dietary fiber, minerals and therapeutic bio-active substances. As per FAO, quinoa is the only staple food for the growing population of the world to provide food security in the next century. In 1993, NASA named quinoa as great food to take it into space as it feeds for a longer period of time with more stabilization of blood sugar preventing fluctuation in energy level.

The declaration of 2013 as "International year of quinoa "by UN Assembly and its growing role in global food security, quinoa attracted people worldwide especially of North-America, France, Australia, China, India and other European and Asian developing countries. As a result, its demand has increased dramatically in recent years to prevent hunger and mal-nutrition around the world. Quinoa, thus could meet the challenges of hunger led malnutrition. In spite of its wide adaptability, rusticity and nutritional superiority, quinoa remains unfamiliar to majority of people in our country and its commercial potential has remained unexplored.

8. Market potential:

Quinoa grains are consumed directly like rice and also in preparation of different food items in India and most of the countries. It is also used as a basic raw material in making breads, cookies, cakes and other bakery products. Since, wheat ATTA is rich in harmful gluten, deficient in essential amino acids, fatty acids, minerals and fibres, it cannot fulfil our nutritional requirements, but makes various kinds of harms and digestive disorders in the body like deterioration of colon and intestinal health, prevention of absorption etc.

Preparation of food items from quinoa grains is a new concept. Its demand is increasing sky-rocketing as it is rich in secondary metabolites and it is fit for every healthy people and also who suffering from diabetes, blood-pressure, arthritis, heart-failure, intestinal disorders, cancer, obesity, celiac, pre-mature hair graying and hair loss, skin problems, sexual weakness and general weakness. About 3, 25,000 tons quinoa grains are needed annually to feed present Indian population. Like this, other countries are also needed it for healthy life in million tons. Thus, processed quinoa grains has unlimited demand and scope worldover for making either Chapattis or various kinds of food products avoiding harmful wheat Atta.

9. Availability of Quinoa grains for processing:

With a view to ensure food grain production, eliminate poverty, hunger and malnutrition and sustain economic development in more dominant marginal areas of India, an Integrated Quinoa Development Project (IQDP) was initiated during September, 2009 in 75 potential districts of 8 States in India by Indian medicinal Plants Marketing Federation, New Delhi. Under this programme, the quinoa presently is being cultivated on about 6278 hectares with total production of 153811 tons per year (As per Annual Report 2015–16, IMPMF, New Delhi). Before consumption of quinoa grains, processing is essential to remove saponin present on the grains which is harmful.

Under this situation, sufficient raw material i.e., 36000 tons needed quantity as per capacity of unit, will be available to run unit year round for processing.

10. Assumptions:

- Processing capacity of unit will be 100 tons per day. Unit will run for 300 days each year.
- Requirement of quinoa grains for processing will be 30000 tons per year.
- Processing charges will be Rs. 30/- per Kg.
- Labour wages have been taken as par market rates.
- The rate of interest has been taken 14 percent on an average both for fixed and working capital.
- The rate quoted in respect of machineries / equipment and raw materials are those prevailing at the time of preparation of report band are likely to vary from place to place and supplier, to supplier, and necessary changes are to be made as and when required.
- Power requirement of the unit is estimated to be 300 HP.
- There will be no problem of pollution due to unit except the dust which may damage the plant and machineries and adversely will affect the health of labour. For this, cyclones may be used to separate the dust.

11. Implementation schedule:

The appropriate time required for various activities is given bellow:

- Selection of site : One month.
- FSSAI Registration / license : 15 days.
- Project report preparation : One month.
- Finance availability : 2 Months
- Machineries procurement, erection, trial etc. : 5 Months
- Construction, erection, commissioning etc. : 8 Months.

12. Technical aspect:

- Quinoa grains will be cleaned thoroughly to remove dust, stones, and other foreign matters before grinding.
- The grains will be processed to remove saponin from the grains for direct consumption.
- The saponin will be utilized in preparation of bio-pesticide.

13. Financial aspect:

A. Fixed capital

Land & Building:

- Land purchase (5 Acres @ Rs. 40 lakh / Acres) : Rs. 20000000/-

- Construction / covered area-3000 Sq.M. for office : Rs. 7500000/-
- Storage etc. (@ Rs. 2500/- per Sq.M.)
- Buildup area for plant & machineries – 2000 Sq. : Rs. 3000000/- (@ Rs. 1500/- per Sq.M.)
- Total : Rs. 2, 05, 00,000/-

Machineries & equipment: Rs. 6, 50, 00,000/-

- ***Erection & consultancy charges:*** Rs. 5,00,000/-
- ***Office furniture & fixture:*** Rs. 5,00,000/-
- ***Pre-operative expenses*** : Rs. 5,00,000/-
- **Total fixed capital (A):** Rs. 10, 05, 00,000/-

B. Working capital (Per month)

Working personnel (Per month):

S.NO.	Designation	No.	Salary per month (Rs.)	Total (Rs.)
1	Manager	1	25000/-	25000/-
2	Chemist (quality control)	1	20000/-	20000/-
3	Electrician	1	15000/-	15000/-
4	Plant Foreman	1	16000/-	16000/-
5	Operator	3	12000/- each	36000/-
6	Accountant	1	15000/-	15000/-
7	Clerk / store keeper	2	12000/-Each	24000/-
8	Un-skilled workers	15	9000/- Each	135000/-
9	Sales supervisors	2	12500/- Each	25000/-
10	Peon	5	8500/- Each	42500/-
11	Security personnel	1	10000/-	10000/-
	Total	**33**	-	**203500/-**
	Add pre-requisites @ 10% of salary	-	-	20350/-
	Total	-	-	**2,23,850/-**

Raw material (Per month):

- Requirement of quinoa grains per month - 3000 Tons.
- Value of grains per month (Average @ Rs. 450/- per Kg - Rs. 135/-Crores
- No. of packing bags required (each of 1 Kg) – 30 lakh
- Value of packing bags (@ Rs. 15/- each) (15 x 300000) - Rs. 4500000/-
- **Total - Rs. 4500000/-**

Utilities (Per month):

- Power (Estimated) - Rs. 10,00,000/-
- Water expenses (Estimated) - Rs. 1,00,000/-
- Total - Rs. 11, 00,000/-

Other contingent expenses:

- Stationary & postage - Rs. 20000/-
- Consumable items - Rs. 100000/-
- Repair and maintenance - Rs. 500000/-
- Advertisement and publicity - Rs. 1000000/-
- Transportation expenses - Rs. 500000/-
- Other miscellaneous expenses - Rs. 500000/-
- Total - Rs. 26, 20,000/-
- Total working Capital (B) - Rs. 8443850/-
- ***Working capital (per month)* - Rs. 84,43,850/-**
- ***Working capital for 2 months* - Rs. 1,68,87,700/-**

Total capital investment:

- Fixed capital - Rs. 10,05,00,000/-
- Working capital for 2 months - Rs. 1,68,87,700/-
- Total - 11, 73, 87,700/-
- Or Say- Rs. 11, 73, 88,000/-

C. Financial analysis

a. Expenses on processing (per year):

- Total Recurring cost per year : Rs. 202652400/-
- Depreciation on building (@ 5% per annum) : Rs. 1, 01,32,620/-
- Depreciation on machineries (@ 10%) : Rs. 65,00,000/-
- Depreciation on furniture & fixed assets (@20%) : Rs. 1,00,000/-
- Interest on total investment (@ 14%) : Rs. 1,64,34,320/-
- **Total : Rs. 23, 58, 19,340/-**
- Or say : Rs. 23.60 Crores

b. Income per Year:

By processing of 30000 tons grains **:** Rs. 90 Crores
(@ Rs. 30/- per Kg)
(After 25% wastage on 36000 tone)

c. *Net profit per year (B – A) : Rs. 66.40 Crores*

Profit – I Year - Rs. 66.40 Crores

After I years - Rs. 88.31 Crores

ANNEXURE - XI

Important suppliers of grains / value added products of Quinoa in India

S.No.	Name of suppliers	State / Location
1	Go Bharaari Agro Industries & Sons Vigyanpuri, Hyderabad	Andhra Pradesh
2	Cerio Export Pvt. Ltd, Jubilee Hills, Hyderabad	
3	Aswartha Traders, Main Bazar, Anantapur	
4	Rich Millet Foods Private Limited, Moula Ali, Hyderabad	
5	Kilaru Naturals Pvt Ltd., Somajiguda, Hyderabad.	
6	Aswartha Traders, Main Bazar, Anantapur	
7	Climate growth Solutions, Begur, Anantapur	
8	Inner Being Wellness Pvt. Ltd Panjagutta, Hyderabad	
9	Quinoa nutritiions Pvt. ltd, Kanchar Bazar, Hyderabad	
10	Vie Foods Private Limited, Hyderabad	
11	Dakshaini Aroma Industries, Anantapur	
12	Nutricircle Limited, Ayesha Residency, Hyderabad	
13	Jeet Enterprises, Fatehnagar, New Delhi	Delhi
14	Kushal Export & Marketing Pvt. Ltd, Lajpatnagar, New Delhi	
15	Krishna Trading Company, Khari Baoli, Delhi	
16	Kenny Delights, Azadpur, Delhi	
17	Mahadev Enterprises Khari Baoli, Delhi	
18	Egon Exim LLP, Sola, Ahmedabad	Gujarat
19	Herbs & Crops Overseas, Chamanpura, Ahmedabad	
20	Mohit Agency, Ahmedabad	
21	Pipaliya Flax Seed, Kuwadva Road, Rajkot	
22	Sahkar Exports, Kalanala, Bhavnagar	
23	Biohack Science Private Limited, Sector-21 C, Faridabad	Harynana
24	Global Organics, IMT Manesar, Gurugram	

25	Fresh Calm Foods & supplements, Karnool, Nandyal	Karnataka
26	Nutriplanet Foods Private Limited, HSR Layout, Bengaluru	
27	Absolute Organic, Whitefield, Bengaluru	
28	Amazing Enterprises, Bengaluru	
29	Orga Foods, Coimbatoor	
30	The Herbs N Spices, Industrial Estate, Neemuch	Madhya Pradesh
31	Trifid Agro Industries, Vijainagar, Indore	
32	Atharva Agro Organics, City Centre, Indore	
33	Shreeveer Organics Agro India, Neemuch Tehsil, Neemuchh	
34	RR Agro Foods, Neemuch	
35	J.C.Enterprise, Malad West, Mumbai	Maharashtra
36	Food Cognicstrading & Consultancy, Vishalnagar, Pune	
37	Happy Earth Life Style , Pvt. Ltd, Marol, Mumbai	
38	Health Happiness, Worli, Mumbai	
39	Palan Traders, Mulund West, Mumbai	
40	Samriddhi Organic Farm (India) Hadapsar, Pune	
41	Bargi International, Saraswathpuram, Mysore	Mysore
42	PCM Export, Sahidnagar, Bhuwaneshwar	Odisha
43	Astha Traders, Patiala	Punjab
44	Hi Tech Agro Services Corporation, Power House Road, Bhatinda	
45	Anshika Lifestyle and Wellness, Heerawatnagar, Jaipur	Rajasthan
46	Apex International C Schemes, Jaipur	
47	PNR Exim, Sayla, Jalore	
48	Queens quinoa processors LLP, Durga Nursery Road, Udaipur	
49	MI Life Style Marketing Private Limited, Malviya Nagar, Jaipur	
50	Smart Technoogies, Azadnagar, Bhilwara	
51	Pankaj Traders, Industrial area, Jalore	
52	PNH Traders, Jodhpur	
53	Rapid Organic Pvt. Ltd, Narnada Colony, Sanchor	
54	Vedikka Enterprises, Viluppuram, Chennai	Tamil Nadu

55	K.V.Associate, Pinchhore, Jhansi	Uttar Pradesh
56	Shivaay Trading Company, Jain Nagar, Merut	
57	Milifestyle Markting Private Limite, Slarpur, Noida	
58	The Taj Urban Grains, Govindpuram, Ghaziabad	
59	Archna Enterprisses, Chilli, Rath	
60	NNAyurvedic Aushadhiya Associate Nishatganj, Lucknow	
61	Organic India, Gomatinagar, Lucknow	
62	Pure & Sure International, Sonapur, Kolkata	West Bengal

ANNEXURE - XII

Important Manufactures / Suppliers of Quinoa seed / grains processing machines in India

S.No.	Name and Address	Type of Machines
1.	Ang Enterprise, No.747 / 9, GIDC Estate, Makarpura, Vadodara, Gujarat	Seeds & Grains processing machines
2.	Goldin (India) Equipments Private Ltd, F-29, BIDC Industrial Estate, Gorwa, Near- Amarkar Maruti Showroom, Badodara, Gujarat	Seeds & Grains processing machines
3.	Rajani Metal Works, Dhobighat, Ahemdabad, Gujarat	Rotary Grain Drum Sleev Machine,
4.	Chintan Industries Amaraiwadi, Ahemdabad, Gujarat	Grain Winnowers,
5.	Nilkant Agri-tech, Dadabhagwan Complex, Surat, Gujarat	Rotary seed cleaner,
6.	Shree Laxmi Vijai Iron & Brass Enterprise, Highway, Sidhpur	Grain processing machine
7.	Nunes Instruments, Avinashi Roa, Coimbatore, Karnataka	Grain polisher machine
8.	Flour Tech Engineers Pvt. Ltd, Plot No. 182, Industrial Area, Sector - 24, Faridabad, Haryana	Grain pre-cleaner
9.	Vrindavan Udyog (India), Connaut place, New Delhi	Grain processing machine
10.	Bharat Agro-Tech Industries, Uppal Southend, Gurgaon, Haryana	Grain processing machine
11.	Agro Asian Industries, Plot No. 135, HSIIDC Industrial Growth Center, Saha, Ambala Cantt, Ambala, Haryana	Grain pre - cleaner machine
12.	Shree Ram Agro Industries, Village-Manglai, PO-Khudda Kalan, Ambala Jagadhary Road, Distt. - Ambala, Haryana	Seed pre- cleaner machine
13.	Hi-tech Agro Industries, G-1, 253D, Road No.12, BehindShri Shyam Dharam Kanta, VKI Area, Jaipur, Rajasthan, India	Seed Grading machine
14.	Kirpa Agro Industries, Plot No. 218, Growth Center, HSIIDC, Industrial Area, Ambala Cantt, Ambala, Haryana	Seed Grading machine
15.	Stitch Experts India, F-94, Mayapuri Industrial Area, Phase-II, New Delhi, Delhi, India	Seed Grading machine

PHOTO GALLARY

SEMINAR AND CONFERENCES ON QUINOA

Fig. 55: *Field day of Quinoa at Sangamner (Maharashtra) organised by "Sarvati Devi Memorial Trust" under "Super foods for Nutritional Food Security and Doubling Farm-income (SNFDFI)" programme*

Fig. 56: *State-level conferance organised by Government of Uttar Pradesh at Lucknow on Production of Super food in dryland areas for livelihood generation*

Fig. 57: *Meeting organised by the Secretary, Irrigation Department, Govt. of Uttar Pradesh at Lucknow on Quinoa development in Bundelkhand Region of U.P.*

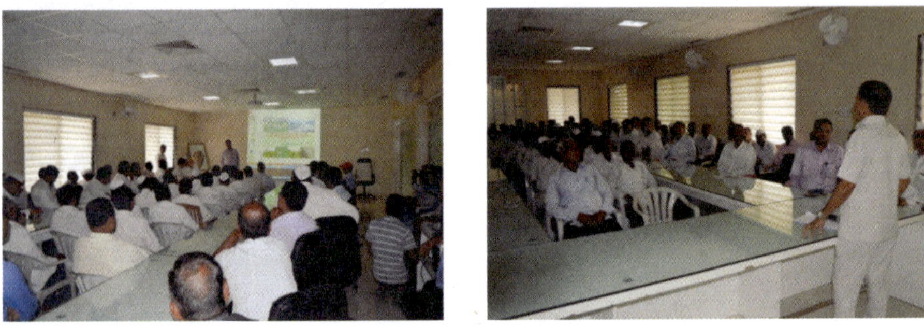

Fig. 58: *National Conferance on "Quinoa development in Maharashtra" organised by "Sarvati Devi Memorial Trust" at Ahmednagar under "Super foods for Nutritional Food Security and Doubling Farm-income (SNFDFI)" programme*

Fig. 59: *National seminar on "Threats of Chemical farming and role of Quinoa in health security" organised by "Sarvati Devi Memorial Trust" at Ahmednagar District of Maharashtra under "Super foods for Nutritional Food Security and Doubling Farm-income (SNFDFI)" programme*

Fig. 60: *National conferance on "Role of Quinoa in health security and livelihood generation" organised by "Sarvati Devi Memorial Trust" at Sangamner (Maharashtra) under "Super foods for Nutritional Food Security and Doubling Farm-income (SNFDFI)" programme*

Fig. 61: *District-level Farmers meeting on "Doubling farm income through Quinoa" organised by 3D Industries Pvt. Ltd at Jodhpur (Rajasthan) under "Extension Research Systems for Quinoa Technology Transfer" programme.*

Fig. 62: *District-level Farmers meeting on "Doubling farm income through Quinoa production" organised by 3D Industries Pvt. Ltd at Mewat, Faridabad (Haryana) under "Extension Research Systems for Quinoa Technology Transfer" programme.*

Fig. 63: *International conferance on Quinoa - 2016 at Dubbai (UAE) organised by FAO and ICBA*

Fig. 64: *Farmers fair organised by Department of Agriculture, Govt. of Uttar Pradesh at Auraiya on Innovations and Technologial advances in agriculture*

Fig. 65: *Farmers meeting and field visit for site selection for Quinoa seed production organised by "Sarvati Devi Memorial Trust" at Ahmednagar District of Maharashtra*

Fig. 66: *Farmers fair organised by District Administration at Bulandshahar (U.P.) on recent advances in Agriculture and Technological innovations on 23 December, 2016*

Fig. 67: *Farmers fair organised by The Poorvanchal Multi-purpose Multi-state Cooperative society Ltd at Ferozpur (Punjab) on recent technological advances in agriculture*

QUINOA CROP

Fig. 68. *Method of quinoa sowing and ridge making*

Fig. 69: *Germinating crop of quinoa at Sangamner, Maharashtra*

Fig. 70: *Quinoa crop at Regional Research Station of Sarvati Devi Memorial Trust at Sangamner (Maharashtra)*

Fig. 71 (a): *Panicle of Sarvati 5-7) variety*

Fig. 71 (b): *Panicles of Sarvati (S-7) variety*

Fig. 72: *Rouging in quinoa seed crop at Regional Research Station of Sarvati Devi Memorial Trust at Sangamner (Maharashtra)*

Fig. 73 (a): *Bumper C crop of quinoa*

Fig. 73 (b): *Bumper crop of quinoa at different places in India*

TRAINING AND VISITS PROGRAMME OF QUINOA GROWERS AND EXPERT / SCIENTISTS / GOVT. OFFICERS

Fig. 74: *Quinoa crop visited by progressive farmers at different places in India*